TRANSFORM YO
MENTO

THE POWER OF MENTOR

THE MENTORSHIP BLUEPRINT: DESIGN YOUR PATH TO ACHIEVING EXCELLENCE

VOLUME – I

SREEKANTH GANESHI

Books By This Author

The Ultimate Leadership in You

"LEADERSHIP IS NOT ABOUT BEING THE BEST. LEADERSHIP IS ABOUT MAKING EVERYONE ELSE BETTER."

Copyright © Sreekanth Ganeshi 2023

All Rights Reserved.

ISBN: 9798856294766

This book has been published with all reasonable efforts taken to make the material error-free after the consent of the author. No part of this book shall be used or reproduced in any manner whatsoever without written permission from the author, except in the case of brief quotations embodied in critical articles and reviews.

The views expressed in this book are solely those of the author and should not be considered as expert instructions or commands. Each reader is responsible for their own actions and decisions.

The Author of this book is solely responsible and liable for its content including but not limited to the views, representations, descriptions, statements, information, opinions, and references ["Content"]. The Content of this book shall not constitute or be construed or deemed to reflect the opinion or expression of the Publisher or Editor. Neither the Publisher nor Editor endorse or approve the Content of this book or guarantee the reliability, accuracy or completeness of the content published herein and do not make any representations or warranties of any kind, express or implied, including but not limited to the implied warranties of merchantability, fitness for a particular purpose. The Publisher and Editor shall not be liable whatsoever for any errors, or omissions, whether such errors or omissions result from negligence, accident, or any other cause or claims for loss or damages of any kind, including without limitation, indirect or consequential loss or damage arising out of use, inability to use, or about the reliability, accuracy or sufficiency of the information contained in this book.

DEDICATION

This book is a heartfelt tribute to my beloved wife, Seema Ganeshi. Her unwavering support and boundless encouragement have been the driving force behind its creation. Throughout this journey, she has stood by my side, making countless sacrifices and unwavering compromises to ensure the success of this endeavour. Without her by my side, this book would not have come to fruition. Seema, your love, and dedication have been the ultimate inspiration, and I am forever grateful for your presence in my life.

TABLE OF CONTENTS

DEDICATION i

TABLE OF CONTENTS iii

ACKNOWLEDGEMENTS v

PREFACE vii

INTRODUCTION x

CHAPTER 1: THE IMPORTANCE OF MENTORS: HOW THEY CAN TRANSFORM YOUR LIFE 1

CHAPTER 2: FINDING THE RIGHT MENTOR: HOW TO IDENTIFY AND CONNECT WITH THE BEST MENTOR FOR YOU 26

CHAPTER 3: LEARNING FROM MENTORS: STRATEGIES AND TECHNIQUES FOR MAXIMIZING THE BENEFITS OF MENTORSHIP 66

CHAPTER 4: MENTORS IN THE REAL WORLD: CASE STUDIES AND EXAMPLES OF SUCCESSFUL MENTOR-MENTEE RELATIONSHIPS 95

CHAPTER 5: BECOMING A MENTOR: HOW TO PAY IT FORWARD AND HELP OTHERS ACHIEVE THEIR GOALS 120

CHAPTER 6: THE PSYCHOLOGY OF MENTORSHIP: UNDERSTANDING THE DYNAMICS OF MENTOR-MENTEE RELATIONSHIPS 142

To be continued in "The Power of Mentor" Volume-II 167

What is Covered in The Power of Mentor - Volume II 168

Chapter 7: Overcoming Mentorship Challenges: How to Navigate Conflicts and Overcome Obstacles in Mentorship ..168

Chapter 8: The Power of Reverse Mentoring: How Learning from Younger Mentors Can Benefit You 168

Chapter 9: Building a Stronger Network: How Mentorship Can Help You Expand Your Professional and Personal Connections 169

Chapter 10: The Role of Mentors in Leadership Development: How Mentorship Can Help You Become a Better Leader......169

Chapter 11: The Impact of Technology on Mentorship: How to Leverage Digital Tools and Platforms for Effective Mentorship 169

Chapter 12: Mentorship and Diversity: How to Find and Engage Mentors from Different Backgrounds and Perspectives 170

Chapter 13: Mentorship Beyond Boundaries: How to Establish and Maintain Long-Distance Mentor-Mentee Relationships .170

May I ask you for a small favor? 172

ACKNOWLEDGEMENTS

This book is an embodiment of the profound love and encouragement bestowed upon me by my cherished family and friends. First and foremost, I humbly express my gratitude to the divine power that guided me throughout this journey and led me to the place I stand today. Thank you, God, for granting me the strength and inspiration to pen these words.

To my exceptional **parents**, your unwavering support has been the cornerstone of my success. I am forever indebted to you for instilling in me the drive to achieve greatness. My heartfelt thanks to my younger sisters, **Jyothi and Aruna,** and my brother **Kiran Kumar,** whose sacrifices have paved the way for my triumphs. It is through the unwavering support of my family that I have realized my dreams.

A special note of appreciation goes to my soulmate, **Seema Ganeshi**. Your unyielding support in this literary endeavour has been nothing short of remarkable. Your belief in my aspirations has been the fuel that powered this journey.

I am also deeply grateful to my mentors, **Bijay Kumar Khandal and Geetika Khandal**, for their transformative guidance that reshaped my thinking. A heartfelt thank you to my The Power of Gratitude mentor, **Shobha Devraj**, for her wisdom and encouragement. Furthermore, I extend my sincere appreciation to **Som Bathla**, my book publishing mentor, whose expertise guided me through the process of writing and publishing. Also, my career banding mentor **Sakshi Chandraakar.**

Lastly, I extend my heartfelt thanks to you, dear reader. Your decision to pick up and peruse this book is a testament to your passion for personal growth and development. It is my earnest hope that these pages inspire and empower you on your journey of transformational leadership and mentorship.

Thank you from the depths of my heart for being a part of this profound adventure. Your support and engagement are the true sources of motivation that keep the essence of this book alive.

PREFACE

In the pages that follow, you will embark on an extraordinary odyssey through the realm of mentorship - a journey enriched by meticulous research, heartfelt testimonials, and real-world experiences. "The Power of Mentor" is not merely a compilation of theories; it is a profound exploration of the human spirit, illuminated by the transformative force of mentorship.

Drawing from a wealth of knowledge and expertise, this book brings to light the importance of mentors and their unparalleled ability to reshape destinies. Our quest for understanding has led us to dive deep into the realms of neuroscience and psychology, revealing the intricate workings of mentorship and its profound effects on the human brain and behaviour.

But beyond the scientific foundation, we have sought to capture the essence of mentorship through the poignant stories of those who have been touched by its magic. Real-world experiences and testimonials from individuals across diverse fields have breathed life into these pages, vividly illustrating how mentors have guided them to new heights, igniting the spark of inspiration that forever changed their lives.

The voices of these mentors and mentees resonate throughout the book, sharing the invaluable lessons and wisdom that have shaped their journeys. Through these authentic narratives, we bear witness to the extraordinary power of mentorship in fostering personal and professional growth, transforming mere aspirations into resounding achievements.

As you traverse the chapters of this book, you will discover practical insights into finding the right mentor, establishing meaningful connections, and cultivating trust and rapport. Through vivid case studies and inspiring examples, you will be inspired by the indelible impact mentors have had on celebrities, corporate leaders, educators, athletes, and those making a difference in their communities through non-profit initiatives.

We recognize that mentorship is a dynamic relationship, and the wisdom shared here extends not only to those seeking mentors but also to those eager to become mentors themselves. Our exploration of the psychology of mentorship lays bares the intricacies of this symbiotic connection, enriching both mentor and mentee, and shedding light on how to overcome challenges that may arise on this shared journey.

This book also delves into the vital role of technology in modern mentorship, bridging gaps and transcending boundaries to connect mentors and mentees from all corners of the world. It embraces the importance of diversity, understanding that unique perspectives fuel innovation and create a tapestry of inspiration that knows no bounds.

"The Power of Mentor" is more than just words on paper; it is a testament to the profound impact mentorship can have on our lives and the lives of those we touch. It is a celebration of the indomitable human spirit and the potential we can unlock when guided by the wisdom and care of a mentor.

With each turn of the page, we invite you to embrace the transformative power of mentorship and to be inspired by the journeys of those who have walked this path before. May the wisdom shared within these pages ignite a passion for mentorship and may the stories of mentor-mentee relationships kindle a desire to create a better world, one connection at a time.

Welcome to "The Power of Mentor" - a journey that will inspire, uplift, and empower you to reach new heights and unleash the greatness within.

INTRODUCTION

With unwavering determination, I set out to write this book, driven by a profound purpose to share the wisdom I've gathered from the crucibles of life. It fills me with profound joy to present my ideas and learnings, eager to connect with you and impart knowledge that could ignite transformative change in your life.

Becoming an author wasn't a lifelong dream; the seed was planted unexpectedly when I stumbled upon a book authored by my dear friend Rupali Goyal. Her name on the cover left me spellbound, and a dormant aspiration ignited within me – the desire to become an "Author" myself.

As a child, I shied away from books, hardly cherishing the world of knowledge they held. Even after graduating, the pages of books remained largely untouched in my life. But fate has a way of guiding us towards unforeseen paths. A sudden twist of events shook the foundation of my corporate life, leaving me without a job and facing an uncertain future. Simultaneously, my beloved sister passed away, leaving behind two young children. In an instant, the responsibility to care for these innocent souls fell upon my shoulders. Unaware of their mother's departure, their grief-stricken faces pulled at my heartstrings.

Amid unimaginable loss and heartache, I faced daunting financial commitments, yet no job in hand to support them. My world seemed to crumble around me, and I was left hoping for everything to be fine. But amidst the shadows of adversity, a glimmer of hope began to emerge. I encountered a remarkable mentor who ushered me into a community that transformed the trajectory of my life.

Guided by this mentor's wisdom, I embarked on a journey of voracious reading and profound self-discovery. It altered my very essence and changed the course of my life yet again. I discovered newfound resilience and an unyielding spirit within the depths of struggle. Through countless nights of uncertainty, I held onto the belief that someday, the tide would turn.

With each page I turned, I felt myself growing stronger, inching closer to my dreams. I navigated the trials of life, shouldering the responsibility of my sister's children, determined to provide them with a future full of love and opportunity.

It was during this tumultuous time that the dream to become an author surged within me. I yearned to create a positive impact in society, to reach out and connect with hearts, and inspire transformative change in countless lives.

This book represents the culmination of that dream – a testament to resilience, hope, and the power of the human spirit. As I stand on the cusp of publishing this book, on the special occasion of my wedding anniversary, August 12th, I am filled with immense gratitude for the journey that brought me here.

I write with the intent to positively impact the lives of 10 million people, to help unlock their boundless potential, and to set them on a path of greatness. I am overjoyed that you have joined me on this remarkable voyage of transformation and growth. Together, let us rise above the challenges, embrace the victories, and create the lives of our dreams.

Thank you, dear reader, for being a part of this extraordinary journey. Your presence here is a testament to your commitment to personal growth and transformation. As we journey onward, united by the pages of this book, let us conquer every obstacle and embrace the boundless possibilities that lie ahead.

CHAPTER 1: THE IMPORTANCE OF MENTORS: HOW THEY CAN TRANSFORM YOUR LIFE

"A mentor is someone who allows you to see the hope inside yourself" – Oprah Winfrey

Have you ever felt stuck in life, not knowing how to move forward to achieve your goals, or things are not moving as you expected? This is where a mentor can help you to find potential and hope inside yourself. This example can help you for a better understanding. Once upon a time, a young man worked at a factory. His mentor, an old technician, taught him to talk less, do more, and never stop developing skills in every aspect of the factory's operation. Ten years later, the old man retired, and the young man became a technician himself. He continued to do his work with the same dedication and diligence as he was taught.

One day, he visited his mentor. The old man saw that he seemed unhappy and asked what was troubling him. The young man sighed and poured his problem out. I have been following your instruction exactly all these years. I kept quiet and focused on the job no matter what I worked on. I know I have done good work at the factory and learned all the skills that can be learned there. What I don't understand is that the guys who don't have my experience and capabilities have all been promoted while I am still making as little as I did before when I was your apprentice. The mentor asked: Are you positive that you have become indispensable to the factory? The young man nodded: Yes. The old man paced and forth to think. After a while, he turned to the young man: You must request a day off, using whatever reason you like. It's time for you to give yourself a break. The young man was surprised by this advice, but the more he thought about it, the more it made sense. He thanked his teacher and left quickly to make a time-off request. When he returned to work after his day off, the manager called him into the office to tell him that things did not go well at the factory while he was gone. Others encountered many problems that were normally handled by him, and they had no idea how to solve them. Realizing his importance, the manager decided to promote him to the position of senior technician, to thank him and encourage him to keep up the good work. The young man was grateful for his mentor's wisdom. Surely, he thought, this was the secret to success. From this point on, whenever the young man felt like he deserved more than what he was getting he would take a day off. When he came back the next day the situation would improve to his satisfaction. This pattern continued for months.

One day the management noticed that this man is taking advantage of the situation and he was blocked from going into the factory. Much to his shock, he found out that his employment was terminated. He could not believe it. Not knowing what else to do, he went back to his mentor to figure out how things went so wrong. Why did I lose my job? he asked with wounded pride. Did I not do everything as you instructed? You did not say the mentor. Because you heard only half the lesson, the old man shook his head. You understood right away that no one pays any attention to a light bulb that is always on, it is only when it goes off that people suddenly take notice and realize they've been taking it for granted. You were so eager to apply this understanding that you left before hearing the second half. Second half? The mentor spoke slowly to make his point" The second half, more important than the first, is the realization that if a light bulb goes off frequently, then sooner or later it will be replaced with one that is more reliable. Who wants a light bulb that no one can count on to provide illumination? You see in life. In your life! Do you have friends and family members that you take for granted, they're always there for you what happens one day they're no longer there do not wait for such a day, to suddenly realize how important they are to give thanks today, for the good fortune of having them in your life also, don't let yourself be taken for granted but at the same time, do your work and don't stop everything you do, just because it doesn't go the way you want it to go find the balance in life and life will reward you with giving you the balance back.

This young technician was struck in his life, he has dedicated himself to his work and abilities to manage all the factory issues to resolve. But things are not moving as expected and he got demotivated. But unknowingly he went to meet his mentor. The mentor noticed his unhappiness and asked the reason, and he discloses his troubles to him. The mentor calmly listened to his every pain and gave the right and wonderful solutions to him. When he applied the mentor's advice got great successful results but listened to only half of the lesson not understanding the impact of another half. When his situation got worse, he eagerly started learning another half of the lesson. If he hadn't met his mentor, he might be thinking that his abilities are useless, and all his efforts are wasted, might get demotivated towards his life, and didn't believe in hard work at all.

1. A mentor can guide and inspire us to reach our potential and overcome challenges in our life.

2. It's important to find a balance in life between work and personal relationships and not take anyone for granted.

3. We should listen to the complete message and not just half of it, as incomplete knowledge can lead to misunderstanding and failure.

4. Hard work and dedication are important, but sometimes taking a break can help us gain perspective and realize our own value and importance.

5. Success is not just about individual achievements but also about the contribution we make to the team and the organization.

The Benefits of Mentorship: Why Everyone Needs a Mentor

Mentors have a vital role to play in our career development. We see mentors as guides and advisors that can help us become the best version of ourselves. No matter where they are in live in life, everyone needs a mentor. There is always a space for growth, development, and progress, regardless of your age or whether you are presently, where you want to be. My life has been greatly impacted by several mentors. Through their wise counsel and direction, I have improved greatly as a person after battling for several years as a lost leader. To achieve an abundant mentality while living with love and purpose, I have learnt everything from strategic career planning, authentic networking, and brand building.

Here is how mentorship helps to be beneficial for everyone

Holds you accountable for your goals: Find a mentor who will encourage, motivate, and challenge you to live beyond your wildest potential. A mentor who will help you set realistic goals that align with your purpose. One who will dive deep and ask your personal and thought-provoking questions that will enhance your focus, clarity, and self-awareness. Your mentor is your accountability buddy; someone who will guide you to stay on track from distractions and limiting beliefs.

Shows you the potential opportunity with appropriate resources and guidance: Mentors are invaluable resources that can provide individuals with the guidance and support needed to reach their full potential. They can show you the potential opportunity with appropriate resources and guidance. Ryan's experience gives you a better understanding of how he achieved success with the help of his mentor.

Ryan had always dreamed of pursuing a career in the music industry, but he wasn't sure where to start. He had little knowledge of the industry and was intimidated by the prospect of breaking in. That's when he met his mentor, a seasoned music executive who took Ryan under his wing. With his mentor's guidance, Ryan was able to navigate the complexities of the music industry and build a successful career. His mentor helped him make connections, learn the ins and outs of the business, and provided him with the resources he needed to succeed. Through their work together, Ryan was able to secure a job at a prominent music label, and he has since gone on to work with some of the biggest names in the industry. Ryan's success is just one example of the potential opportunities that mentors can provide with appropriate resources and guidance. Whether it's navigating a new career path, starting a business, or pursuing a personal passion, mentors can provide the knowledge and support needed to succeed. With their help, individuals can unlock their full potential and achieve their goals, making the impossible possible.

Helps you expand your network: Sharing your vision, accomplishments, and struggles will help you build your relationship with your mentor. Every time you connect with someone in person or online, whether they will become your mentor in a long-term is not important. Through every conversation, you will develop new skills, share knowledge, and learn from different prospectives. Be open and vulnerable in person and online and don't be afraid to ask for help. You'd be surprised who will notice you, offer to help, promote you, or even introduce you to someone in their network. Genuinely ask for introductions and offer to do the same. Your network will expand so will your mastermind group of like-minded, ambitious who share similar visions and mindsets. The power of mentorship is that it's an investment in your personal and professional development and is mutually beneficial for both you and your mentor. Find a mentor who is a role model and inspiration, someone who will invest in your growth and well-being – you will accelerate to greater heights with the help of others. Find a mentor but become one too and pay it forward.

The Science of Mentorship: How It Affects Your Brain and Behavior

Mentorship is a valuable tool for personal and professional growth. Studies have shown that having a mentor can lead to increased job satisfaction, career success, and even greater well-being. However, the science of mentorship goes beyond just these benefits, it also affects our brains and behaviour in powerful ways. One of the keyways mentorships affect our brain is through the process of social learning. Social learning is the process by which we learn from the behaviours and experiences of others. When we have a mentor, we have the opportunity to observe their behaviours, attitudes, and values. Over time, we begin to adopt some of these behaviours, which can influence our own brains and behaviour. Mentorship can also lead to the formation of new neural connections in the brain. Studies have shown that when we engage in activities that challenge us and push us outside of our comfort zone, new neural connections are formed. Mentors can help us identify and pursue these challenging activities, leading to increased neural plasticity and improved cognitive function.

In addition to these brain-related effects, mentorship can also influence our behavior in other ways. For example, having a mentor can lead to increased self-confidence and self-efficacy, which can in turn lead to greater success in various areas of our lives. Mentors can also provide us with valuable feedback and guidance, helping us to identify areas where we can improve and providing us with strategies for making those improvements. Overall, the science of mentorship demonstrates the powerful impact that mentors can have on our brains and behavior. By providing us with new learning opportunities, challenging us to grow and improve, and offering guidance and feedback, mentors can help us achieve our goals and live more fulfilling lives. Here are a few real-time experiences about positive and negative leadership mentor, how it damages our brain and behaviour and organization.

Positive leadership can have a profound impact on emotional well-being, inspiration, relationships, trust, and logical thinking. Leaders who have a positive attitude and exhibit emotional intelligence can help create a work environment where employees feel valued and motivated. Inspirational leaders can spark creativity and innovation among their team members, leading to better problem-solving and increased productivity. Strong relationships between leaders and team members can create a sense of trust and foster open communication, which can lead to better collaboration and a more supportive work environment. An example of positive leadership can be seen in the success of the tech company Google. The company's co-founders, Larry Page and Sergey Brin, created a culture of innovation and creativity by promoting a positive attitude and encouraging employees to take risks. They also created a supportive work environment by offering generous benefits and investing in employee well-being. On the other hand, negative leadership can have a detrimental effect on emotional well-being, inspiration, relationships, trust, and logical thinking. Leaders who exhibit negative behaviours, such as micromanaging, bullying, or showing favouritism, can create a toxic work environment that leads to decreased morale, increased stress, and high turnover rates.

One real-time example of negative leadership can be seen in the downfall of the ride-sharing company Uber. The company's former CEO, Travis Kalancik was accused of fostering a toxic work culture that included sexual harassment and bullying. This behaviour led to decreased morale among employees and a negative public image for the company. The science of mentorship also plays a crucial role in leadership. Mentors can provide guidance, support, and feedback that can help employees develop new skills and advance in their careers. Mentorship can also have a positive effect on the brain and behavior. Studies have shown that mentorship can increase brain activity in areas associated with learning, memory, and emotional regulation. It can also lead to increased confidence and a more positive outlook on the future.

One example of effective mentorship can be seen in the work of basketball coach Gregg Popovich. Popovich has been praised for his ability to develop young players and help them reach their full potential. His mentoring approach emphasizes communication, trust, and support, which has led to the success of his team, the San Antonio

Spurs. Positive leadership can have a powerful impact on emotional well-being, inspiration, relationships, trust, and logical thinking. Conversely, negative leadership can have a detrimental effect on these same areas. Effective mentorship can play a crucial role in developing positive leadership skills and can lead to positive brain and behavioral effects.

The Power of Role Models: Why We Look Up to Mentors

Jyoti Kumari, a 15-year-old girl from Darbhanga, Bihar India made headlines in May-2020 when she cycled 1200 km from Gurugram to Darbhanga, with her father sitting on the pillion, during the nationwide lockdown in India. Her story is a shining example of the power of role models and how they can inspire us to achieve extraordinary feats. Jyoti's father, Mohan Paswan, had injured his leg and was unable to work during the lockdown, which left the family with no income. With no money to pay rent and buy food, Jyoti decided to take matters into her own hands. She had heard about the Shramik special trains, which were ferrying migrant workers back to their homes, but with no money for the train tickets, Jyoti came up with a bold plan. Jyoti decided to cycle all the way from Gurugram to Darbhanga, covering a distance of 1200 Km, with her father sitting on the pillion. She had never cycled such a long distance before, but she was determined to do whatever it took to get her family home. With no proper gear or training, Jyoti and her father set off on their epic journey. Their journey was not without its challenges. They face extreme heat, hunger, exhaustion, and even a flat tire. But Jyoti never lost hope. She pushed herself to the limit, pedaling for hours on end and inspiring her father to keep going. Her story soon went viral, and she became a symbol of hope and determination for millions of people in India.

Jyoti's story is a testament to the power of role models and how they can inspire us to overcome even the toughest challenges. Jyoti was inspired by her mother, who passed away a few years ago. Her mother was a strong and resilient woman who had overcome numerous challenges in her life, and Jyoti looked up to her as her role model. Jyoti's mother's spirit and determination lived on in Jyoti, who refused to give up even in the face of extreme adversity. She drew strength from her mother's memory and showed the world what a young girl from a small town in Bihar can achieve with determination and hard work. Jyoti's story also highlights the importance of mentorship and guidance. She was fortunate enough to have a mentor in her uncle, who taught her how to ride a bike and instilled in her the values of hard work and perseverance. Jyoti's uncle was her role model, and his guidance and support helped her achieve her goals. **Jyoti Kumari was conferred the 2021 "Pradhan Mantri Rashtriya Bal Shakti Puraskar" for bravery in the times of pandemic by the President of India.** Jyoti was invited to appear for cycling trials at the "All India Cycling Federation" because she was a plausible selection for the Indian Cycling team. The New

York Times referred to Jyoti as the "Lionhearted Girl, Inspiring a Nation".

Jyoti Kumari's story is a shining example of the power of role models and how they can inspire us to achieve extraordinary feats. Her story touched the hearts of millions of people in India and around the world, and she became a symbol of hope and determination during the challenging times of the pandemic. Jyoti's story also reminds us of the importance of mentorship and guidance in our lives. Mentors like Jyoti's uncle can have a profound impact on the lives of those they guide, helping them to realize their full potential and achieve their dreams. As we strive to achieve our own goals and overcome the challenges in our lives. We should look to role models like Jyoti Kumari and draw inspiration from their stories. We should also seek out mentors who can guide us, and inspire hard work, and with the support of our role models and mentors, we can achieve anything we set out minds to.

The Role of Mentors in Personal and Professional Growth

Mentors play a crucial role in personal and professional growth, especially in leadership development. A mentor is someone who provides guidance, advice, and support to their mentee in their personal and professional pursuits. Mentors can be found in various fields, including business, education, and sports. They have practical success experiences that can help their mentees achieve their goals and become successful leaders. The role of mentors in leadership development is critical. Effective leadership requires a wide range of skills, including the ability to communicate effectively, make tough decisions, and inspire and motivate others. Mentors can help their mentees to develop these skills by providing guidance and support, offering feedback, and sharing their own experiences and insights. A mentor can help a mentee identify their strengths and weaknesses as a leader. Through regular feedback and constructive criticism, a mentor can help a mentee understand how they are perceived by others and where they can improve. This can be especially valuable in developing self-awareness, a critical trait for effective leadership.

One of the most important ways that mentors can support their mentees is by providing them with opportunities for growth and development. This might involve introducing them to new challenges and experiences, encouraging them to take on leadership roles, or connecting them with other individuals in their field who can help them to grow and develop. A mentor can challenge and push a mentee to grow and improve as a leader. They can encourage mentees to take on new challenges and step outside their comfort zone. This can help mentees develop resilience, a key trait for successful leadership. Mentors can also provide accountability, helping mentees stay on track with their goals and progress. In leadership development, mentors are especially valuable as they can help individuals navigate the complexities of leadership and provide guidance on how to develop the necessary skills and qualities to become an effective leader. A mentor can help mentees identify their strengths and weakness, develop important relationships, offer advice on how to network effectively and provide insights into the culture and norms of their industry. Furthermore, a mentor can serve as a role model for their mentee, demonstrating the qualities and behaviors of effective leadership. By observing their mentor, mentees can learn how to manage their time, prioritize tasks, and make decisions that align with their values and goals.

The Different Types of Mentors: Which One Do You Need?

When seeking a mentor, it's crucial to understand the different types of mentors available and identify which type is best suited for your needs. The type of mentor you choose will depend on your goals and what you hope to gain from the mentor-mentee relationship.

If you are looking for guidance and support on leadership and communication development, a leadership mentor may be the best fit. A leadership mentor can help you identify leadership strengths and weaknesses, offer feedback on leadership and communication style, and provide advice on how to develop effective communication and leadership skills. They can help you understand the qualities of successful leaders and provide insights into how to build trust and inspire others. My mentor Bijay Kumar Khandal expertise in these areas and helped so much in developing my leadership and communication skills.

You can get his mentorship by registering for his master class happens every Tuesday at 8.00.P.M IST link is here https://school.peakimpactmentorship.com/f/work-with-bijay

If you are searching for guidance and support on writing and publishing your own books, the best book writing and self-publishing will be the best fit for you. A self-publishing mentor can help you to identify your topics and offer how to research your content, provide research tools, understand the quality and profitability of your content, and helps you format and publish your book. I am providing you here with my mentor who helped to write and publish my books.

You can also register and self-publish your book in 30 days his name is Som Bathla his master class link is here https://sombathla.com/masterclass

If you are seeking guidance and support on how gratitude transforms your life positively my mentor Shobha Devraj can help you. I have applied those techniques as she suggested it has given me positive results. These types of mentors you can easily trust and build strong relationships with them. She conducts a master class every Tuesday, Wednesday and Saturday at 10 A.M IST

You can mentorship by registering yourself on the "Power of Gratitude Club" with this link https://powerofgratitude.club/

If you are seeking guidance and support on personal and life issues, life mentors may be the best fit. A life mentor can help you identify personal goals, offer advice on how to balance personal and professional life and provide support during challenging times. They can help you navigate personal relationships, deal with personal challenges and provide a sounding board for your ideas.

If you are seeking guidance and support in your career development, a career mentor may be the best fit. A career mentor can help you identify your career goals, provide advice on job search strategies, and offer feedback on your performance. They can help you navigate the job market, offer advice on negotiating salaries, and provide insights into industry-specific skills and knowledge.

I am providing my mentor details who has helped me with career advancement and creating my own brand. If you are the one looking for career advance my mentor Sakshi Chadraakar will help you.

https://brand.sakshichandraakar.com/

If you are seeking guidance and support on industry-specific issues, an industry mentor may be the best fit. An industry mentor can provide insights into industry trends, offer advice on how to navigate industry politics, and offer feedback on industry-related projects. They can help you understand the intricacies of your industry and help you develop the skills and knowledge needed to be successful in your field.

If you are looking to develop technical skills, a technical mentor may be the best fit. A technical mentor can help you learn new skills, identify training opportunities, and provide feedback on technical projects. They can help you develop your technical expertise and provide insights into the latest technological advancements in your field.

KEY TAKEAWAYS

CHAPTER 1: THE IMPORTANCE OF MENTORS: HOW THEY CAN TRANSFORM YOUR LIFE

The Benefits of Mentorship: Why Everyone Needs a Mentor

1. Mentorship holds individuals accountable for their goals and helps them stay focused and motivated.

2. A mentor can provide guidance and support, showing potential opportunities and offering appropriate resources to achieve success.

3. Through mentorship, individuals can expand their network and build valuable relationships with like-minded and ambitious people.

4. Mentorship is an investment in personal and professional development and is mutually beneficial for both the mentee and the mentor.

5. Being open, vulnerable, and willing to ask for help can lead to valuable connections and opportunities for growth.

6. Mentors can offer insights and knowledge from their own experiences, helping mentees navigate challenges and achieve their aspirations.

7. Becoming a mentor, oneself and paying if forward can create a positive impact on others and contribute to the growth of the community as a whole.

Overall, mentorship is a powerful tool for personal and professional growth, providing guidance, support, and valuable connections to help individuals reach their full potential and achieve their goals, By embracing mentorship and being willing to learn from others, individuals can unlock new opportunities and accelerate their journey to success.

The Science Of Mentorship: How It Affects Your Brain and Behavior

1. Mentorship can have profound effects on our brain and behavior, leading to increased job satisfaction, career success, and well-being.

2. Social learning is a key aspect of mentorship, where we adopt behaviors, attitudes, and values from our mentors.

3. Mentorship can lead to the formation of new neural connections in the brain, enhancing cognitive function and neural plasticity.

4. Positive leadership, fostered through mentorship, can create a work environment that promotes emotional well-being, trust, collaboration, and innovation.

5. On the other hand, negative leadership can lead to a toxic work environment, decreased morale, and high turnover rates.

6. Effective mentorship in leadership can positively impact brain activity, confidence, and outlook on the future, leading to better leadership skills and career development.

7. Real-life examples like Google and Uber demonstrate the significant impact of positive and negative leadership on organizations and employee experiences.

Mentorship and Positive leadership go hand, and the science behind it showcases the potential for personal and professional growth when individuals receive guidance, support, and positive role modeling from their mentors. Conversely, negative leadership can have detrimental effects on both individuals and organizations, underscoring the importance of cultivating a positive and supportive work environment through mentorship and effective leadership practices.

The Power Of Role Models: Why We Look Up to Mentors

1. Role models have the power to inspire us to achieve extraordinary feats, even in the face of extreme adversity.

2. Jyoti Kumari's story exemplifies the determination and strength that can be drawn from looking up to role models.
3. Jyoti's mother served as her role model, and her spirit and resilience influenced Jyoti to overcome challenges and pursue her goals.

4. Mentorship and guidance, exemplified by Jyoti's uncle, play a crucial role in helping individuals realize their potential and achieve success.

5. Jyoti's achievements were recognized and celebrated, showcasing the impact of her inspiring story on the nation and beyond.

6. Jyoti's story serves as a reminder that with the support of role models and mentors, anyone can achieve their dreams and overcome obstacles.

7. Drawing inspiration from the experiences of role models like Jyoti can motivate individuals to strive for greatness and persevere through challenging times.

Jyoti Kumari's story is a powerful testament to the influence of role models and mentors in our lives. Her determination and resilience serve as a source of inspiration for millions of people, highlighting the transformative impact that positive role models and supportive mentors can have on individuals' lives and achievements.

The Role of Mentors In Personal And Professional Growth

1. Mentors are instrumental in personal and professional growth, particularly in leadership development, providing valuable guidance and support.

2. Effective leadership requires a diverse skill set, and mentors help mentees develop communication, decision-making, and motivational abilities.

3. Mentors play a key role in helping mentees identify their strengths and weaknesses as leaders, fostering essential self-awareness.

4. Providing opportunities for growth and development is a crucial aspect of mentorship, exposing mentees to new challenges and experiences.

5. Mentors push mentees beyond their comfort zones, fostering resilience, and providing the necessary accountability for progress.

6. In leadership development, mentors offer valuable insights and guidance, helping mentees navigate complex situations and develop crucial skills.

7. Mentors act as role models, exemplifying effective leadership qualities, time management, and decision-making for mentees to emulate.

8. Building important relationships, effective networking, and understanding industry norms and enhanced by mentorship in leadership development.

9. Mentors' practical success experiences enable them to guide mentees effectively toward achieving their goals and becoming successful leaders.

10. Mentors serve as invaluable sources of feedback and constructive criticism, helping mentees improve their leadership capabilities.

Mentorship plays a pivotal role in personal and professional growth, especially in leadership development. The guidance and support mentors provide, along with the opportunities for growth and self-awareness they foster, enable mentees to navigate the complexities of leadership and become effective successful leaders in their respective fields.

The Different Types Of Mentors: Which One Do You Need?

When seeking a mentor, consider the different types available and choose the one aligned with your goals.

1. Leadership Mentor: For guidance on leadership and communication development.

2. Book Writing and Self-Publishing Mentor: To get support with writing and publishing your own books.

3. Gratitude Mentor: For guidance on how gratitude positively transforms your life.

4. Life Mentor: To navigate personal and life issues and find support during challenging times.

5. Career Mentor: For career development, job search strategies, and industry-specific insights.

6. Industry Mentor: To gain insights into industry trends, politics and projects.

7. Technical Mentor: To learn new technical skills and stay updated on technological advancements.

   Selecting the right mentor can be instrumental in achieving personal and professional growth in your chosen field.

CHAPTER 2: FINDING THE RIGHT MENTOR: HOW TO IDENTIFY AND CONNECT WITH THE BEST MENTOR FOR YOU

> *"Finding a mentor is like finding a hidden treasure. It may take some time and effort to uncover, but once you find it, the rewards are invaluable."* –
> Bernard Kelvin Clive, Author, and Personal Branding Coach

Finding the right mentor may take time and effort, but the benefits of having a mentor who can guide and support you in your personal and professional growth can be invaluable.

Identify Your Goals: Start by identifying your goals and what you hope to achieve from a mentorship. This will help you narrow down your search to mentors who have experience and expertise in your area of interest.

Seek Recommendations: Attend networking events, seminars, and industry conferences, and join professional associations or groups related to your industry. These groups often provide mentorship programs or connections with established professionals in your field or reach out to your friends, family, colleagues, or professors who might be able to connect you with potential mentors. You can also use online platforms like LinkedIn or any other social media platforms who are works in your field or have experience in your area of interest.

Research Potential Mentors: Once you have identified potential mentors, do some research to ensure that they have the experience and expertise you are looking for. Look at their professional background, areas of expertise, and past mentorship experiences to determine whether they are the right fit for you. Narrow down your list of potential mentors. Focus on those who align with your goals and objectives, have relevant experience, and are available to provide mentorship.

Reach Out: Once you have identified potential mentors, reach out to them via email or LinkedIn or other social media platforms. Be clear about your goals, what you hope to achieve by working with them, and why you think they would be a good fit for you.

Schedule A Meeting: If the mentor responds positively, schedule a meeting to discuss your goals and expectations. This will give you a chance to learn more about the mentor's experience and expertise and determine whether they are the right fit for you. Be prepared for each meeting or call. Take notes, ask questions, and be open to feedback. Show appreciation for your mentor's time and expertise. Be respectful, punctual and follow through on commitments.

Set Expectations: Set clear expectations for the mentorship. Discuss how often you will meet or communicate, what topics you will cover, and what goals you have to achieve. Take ownership of your own personal growth and development. Be flexible and adaptable. Recognize that your goals and needs may change over time and be open to adjusting your mentorship plan accordingly.

Be Open to Feedback: Remember that mentorship is a two-way street. Be open to feedback and advice from your mentor and use it to improve your skills and achieve your goals. Evaluate your progress regularly. Set milestones and checkpoints to measure your progress towards your goals. Give it back to your mentor. Share your progress and successes with them and acknowledge the impact they have had on your growth and development.

Continue to network and seek out new mentorship opportunities. Keep building relationships with industry professionals and seek out new opportunities for growth and development. Pay it forward. Once you have achieved your goals, consider becoming a mentor yourself and provide guidance and support to others who are starting out in their careers or seeking personal development.

Knowing What You Want: Defining Your Goals and Objectives

Defining your goals and objectives is an essential first step to finding the right mentor for you. Without a clear understanding of what you want to achieve, it can be challenging to identify a mentor who has the expertise, experience and approach that aligns with your goals.

1. Identify your long-term goals: Start by thinking about your overall career or personal aspirations. What do you want to achieve in the next five, ten or twenty years? Write down your goals and be as specific as possible.

2. Break down your long-term goals into smaller objectives: Once you have your long-term goals in mind, break them down into smaller, more manageable objectives. This will help you focus on the steps you need to take to achieve your goals.

3. Prioritize your objectives: Determine which objectives are most important to you and prioritize

them. This will help you stay focused and ensure that you are working towards the goals that matter most to you.

4. Identify the skills and knowledge you need to achieve your objectives: Think about the skills and knowledge you need to achieve your objectives. What areas do you need to develop or improve? This will help you identify the type of mentor you need.

5. Research potential mentors: Once you have a clear understanding of your goals and objectives, start researching potential mentors who can help you achieve them. Look for mentors who have experience in your field or industry, have achieved similar goals, and have a teaching or mentoring style that resonates with you.

6. Reach out to potential mentors: Once you have identified potential mentors, reach out to them to see if they are interested in mentoring you. Be clear about your goals and objectives and why you think they would be a good fit as a mentor.

Building Your Network: How to Meet Potential Mentors

Attend Networking Events: One powerful example of attending networking events is from a marketing professional who attended a marketing conference. At the conference, she made a point of attending as many sessions as possible and introducing herself to other attendees. Through this, she was able to connect with a well-known marketing expert who became her mentor, offering her guidance and support throughout her career.

Use Social Media: Another example comes from a freelance writer who used LinkedIn to connect with potential mentors. By following industry leaders and engaging with their content, she was able to build relationships and eventually found a mentor who provided her with invaluable guidance and connections that helped her grow her business.

Participate in Online Communities: A recent graduate, who was interested in a career in technology, joined an online community for women in technology. Through the community, she was able to connect with women in the industry who provided her with guidance on job searching, networking and professional development.

Attend Informational Interviews: A University student who was interested in a career in law set up informational interviews with several lawyers in the community. Through these interviews, she was able to gain insight into the industry and connect with a mentor who became a valuable resource throughout her education and career.

Volunteer: Another example comes from a software developer who volunteered for a non-profit organization. Through this experience, he was able to work alongside other developers and connect with a mentor who provided him with valuable guidance on career development and technical skills.

Ask for Introductions: A professional in the finance industry asked a former colleague for an introduction to a senior executive at a major financial institution. Through the introduction, she was able to meet the executive and eventually landed a job at the institution, where she continued to build her network and find mentors to support her career growth.

Remember, building your network takes time and effort, but it can be one of the most valuable investments you make in your career. By using these powerful and practical tips and being open to new opportunities, you can find the right mentors to guide you on your journey.

Qualities to Look for in a Mentor: What to Consider When Choosing a Mentor

Openness and Honesty: A mentor who is open and honest with you can help you learn from your mistakes and grow as a person. A mentor who is willing to share their own experiences, challenges, and failures can help you avoid making the same mistakes. I once had a mentor who was very open and honest with me about his own struggles, and his willingness to share his experiences helped me understand that everyone faces challenges and setbacks on their journey to success.

Flexibility and Adaptability: A mentor who is flexible and adaptable can help you navigate the ever-changing landscape of your industry or profession. A mentor who is willing to try new things, experiment with different approaches, and adjust their strategies based on feedback can help you stay ahead of the curve. I once had a mentor who was always willing to try new things and experiment with different strategies, and her flexibility and adaptability helped me develop a growth mindset and become more agile in my approach.

Passion and Enthusiasm: A mentor who is passionate and enthusiastic about their work can help you find your own passion and motivation. A mentor who exudes energy and excitement about their industry or profession can inspire you to pursue your own goals with vigor and enthusiasm. I once had a mentor who was so passionate about her work that it was contagious. Her enthusiasm and energy inspired me to pursue my own dreams and goals with renewed vigor.

Intuition and Emotional Intelligence: A mentor who has a strong sense of intuition and emotional intelligence can help you navigate complex interpersonal situations and build strong relationships. A mentor who is able to read between the lines, understand unspoken cues, and respond with empathy and compassion can help you develop your own emotional intelligence and interpersonal skills. I once had a mentor who had a remarkable ability to read people and situations, and her intuition and emotional intelligence helped me develop a deeper understanding of myself and others.

Strategic Thinking and Vision: A mentor who has a strategic mindset and a clear vision for the future can help you develop your own strategic thinking and planning skills. A mentor who is able to see the big picture, identify trends and opportunities, and develop a plan to achieve long-term goals can help you think beyond the immediate challenges and opportunities in front of you. I once had a mentor who had a clear vision for the future of our industry and was able to guide me in developing my own strategic thinking and planning skills. Her vision and guidance helped me become a more effective leader and achieve my long-term goals.

The Worst Qualities of a Mentor: How They Can Hinder a Mentee's Growth and Development:

Self-Centered: A mentor who is self-centered is primarily interested in their own success and doesn't prioritize the mentee's growth and development. This can lead to a lack of guidance and support for the mentee, who may feel neglected and unimportant. A self-centered mentor may also be more interested in promoting their own brand than in helping the mentee succeed.

Lack of Empathy: A mentor who lacks empathy is unable or unwell to understand the mentee's perspective or feelings. This can make the mentee feel unsupported and demotivated, particularly if they are struggling with personal or professional challenges. A lack of empathy can also make it difficult for the mentor to provide effective guidance and advice.

Dishonesty: A mentor who is dishonest or manipulative can be harmful to the mentee's growth and development. Dishonesty can lead to a breakdown in trust between the mentor and mentee, which can make it difficult for the mentee to rely on the mentor for guidance and support. Additionally, a mentor who encourages unethical or illegal practices can put the mentee's career or reputation at risk.

Control: A mentor who is overly controlling can stifle the mentee's growth and development by preventing them from taking risks or making decisions on their own. This can prevent the mentees from developing their own skills and confidence and can make them overly reliant on the mentor. A controlling mentor may also be resistant to new ideas or approaches, which can limit the mentee's ability to innovate or adapt to changing circumstances.

Lack of Experience: A mentor who lacks experience in the field or industry in which the mentee is seeking guidance may be unable to provide effective guidance or support. A mentor with limited experience may not be able to anticipate the challenges that the mentee will face or may not be familiar with the best practices and strategies in the field. This can make it difficult for the mentee to succeed and can limit their career prospects.

Mentors who exhibit these worst qualities can be detrimental to a mentee's growth and development. When choosing a mentor, it is important to look for someone who is supportive, empathetic, honest, open to new ideas and experienced in the field or industry in which the mentee is seeking guidance.

Approaching Your Mentor: How to Initiate the Mentorship Relationship:

1. **Networking Event:** At a networking event for entrepreneurs, Smita met Rohit, a successful entrepreneur who has founded several companies. Smita was impressed with Rohit's knowledge and experience and asked if he would be willing to mentor her. Initially, Rohit was hesitant, but Smita was persistent and demonstrated her enthusiasm and

dedication to her business. Eventually, Rohit agreed to mentor her and provided valuable insights and advice that helped Smita grow her business.

2. **Email Introduction:** Arjun was a recent graduate who was interested in pursuing a career in technology. He sent an email to Anu, a well-respected technology executive, introducing himself and expressing his interest in a mentorship relationship. Anu was initially tentative because of her busy schedule, but Arjun was consistent and showed his dedication to his career. Eventually, Anu agreed to mentor him and provided guidance and support that helped Arjun launch his career in technology.

3. **Social Media Outreach:** Divya was a young professional who was interested in learning more about marketing. She reached out to Raj, a well-respected digital marketing expert, on LinkedIn and expressed her interest in a mentorship relationship. Raj was initially skeptical because of the large number of mentorships requests he received on social media, but Divya was persistent and demonstrated her enthusiasm for digital marketing. Eventually, Raj agreed to mentor her and provided valuable insights and advice that helped Divya develop her skills in digital marketing.

4. **Referral from Colleagues:** Manish was a recent college graduate who was interested in pursuing a career in finance. A colleague referred him to Shreya,

a successful finance executive who had worked for several top companies. Manish reached out to Shreya and expressed his interest in a mentorship relationship. Shreya was initially hesitant because of her busy schedule, but Manish was persistent and demonstrated his dedication to his career. Finally, Shreya agreed to mentor him and provided guidance and support that helped Manish launch his career in finance.

5. **In-person Request:** Naina was a young professional who was interested in learning more about sales. She approached Ravi, a successful sales executive, at a trade show and asked if he would be willing to mentor her. Ravi was initially hesitant because he did not have much experience with mentorship, but Naina was persistent and demonstrated her enthusiasm for sales. Eventually, Ravi agreed to mentor her and provided valuable insights ad advice that helped Naina developed her skills in sales.

6. **College Alumni Network:** Rakesh was a recent graduate who was interested in pursuing a career in marketing. He reached out to an alumni network and connected with Priya, a successful marketing executive who has graduated from his college. Rakesh expressed his dedication to his career relationship, and Priya was impressed with his dedication to his career. Priya agreed to mentor him and provided guidance and support that helped Rakesh launch his career in marketing.

7. **Professional Organization Meeting:** Sandhya was a young professional who was interested in learning more about leadership. She attended a professional organization meeting and met with Rahul, a well-respected leadership coach. Sandhya expressed her interest in a mentorship relationship, and Rahul was initially hesitant because he had a busy schedule. However, Sandhya was persistent and demonstrated her enthusiasm for leadership. Eventually, Rahul agreed to mentor her and provided valuable insights and advice that helped Sandhya develop her leadership skills.

8. **Direct approach:** Avinash, a young professional in the finance industry, was looking to advance his career and knew that he needed guidance from an experienced mentor. He had been following the work of Ramesh, a well-respected professional in India, and believed that Ramesh would be the perfect mentor for him.

Avinash decided to take a direct approach and reached out to Ramesh, introducing himself and explaining why he wanted to be mentored by him. Ramesh was initially hesitant to take on another mentee, as he already had a busy schedule, but he was impressed by Avinash's dedication and enthusiasm.

Avinash continued to follow up with Ramesh regularly, sending updates on his progress and asking for feedback. He also attended several events where Ramesh was speaking and made sure to introduce himself and express his gratitude for the opportunity to be mentored. Over time, Ramesh became more and more invested in Avinash's success and their mentorship relationship grew stronger. Avinash was able to learn valuable insights and strategies from Ramesh which helped him advance his career and achieve his goals.

In this case, the direct approach may not have initially yielded the desired responses, but Avinash's consistency and dedication ultimately won over his mentor and led to a successful mentorship relationship.

Building Trust and Rapport: How to Develop a Strong Mentor-Mentee Relationship

Building trust and rapport is essential for developing a strong mentor-mentee relationship. Trust is built over time through open communication, honesty, reliability, and mutual respect. A good mentor is someone who can provide guidance, support, and encouragement to their mentee, while also challenging them to grow and develop. Developing a strong relationship requires both parties to be committed to the process and willing to invest time and effort into the relationship. To better understand how to build trust and rapport in a mentor-mentee relationship, a case study is here.

This incident happened in Bangalore.

- Mentor: Kiran, as a seasoned businessman in his late 50s from Mumbai
- Mentee: Anjali, a young entrepreneur in her mid-20s from Bangalore

1. **Initial Misunderstanding**

   Situation: Anjali has just started her business and is looking for guidance from Kiran, who has offered to be her mentor. They meet at a café in Bangalore to discuss their goals and expectations.

   Problem: Kiran starts the conversation by asking Anjali about her family background and education. Anjali feels offended, as she thinks Kiran is questioning her qualifications and abilities.

   Pain: Anjali feels disrespected and loses trust in Kiran. She starts to question if he is the right mentor for her.

   Solution: Kiran realizes his mistake and apologizes to Anjali for his insensitive remark. He assures her that he has full confidence in her abilities and wants to focus on helping her achieve her business goals. Anjali appreciates his apology and agrees to move forward with the mentorship.

2. **Communications Issues**

Situation: Anjali and Kiran have been meeting regularly, but they are struggling to communicate effectively.

Problem: Kiran is used to a more direct communication style, while Anjali tends to be more indirect in her approach. This leads to misunderstandings and frustrations on both sides.

Pain. Anjali feels like Kiran is not understanding her perspective, while Kiran feels like Anjali is not clear about her goals and challenges.

Solution: Kiran and Anjali discuss their communication styles and agree to be more mindful of how they communicate with each other. Kiran also suggests that they set specific goals for their meeting and prepare an agenda beforehand to ensure they stay on track.

3. **Cultural Differences**

Situation: Anjali and Kiran come from different parts of India and have different cultural backgrounds

Problem: Anjali feels like Kiran doesn't understand the unique challenges she faces as a young entrepreneur in Bangalore. Kiran, on the other hand, feels like Anjali is not taking his advice seriously.

Pain: Anjali feels like Kiran is dismissive of her cultural identity, while Kiran feels like Anjali is not open to his perspective.

Solution: Kiran and Anjali discuss their cultural differences and how they can work together to bridge the gap. Kiran suggests that they learn more about each other's cultures and perspectives to better understand each other. Anjali agrees and starts to share more about her experiences as an entrepreneur in Bangalore.

4. **Cultural Sensitivity**

    Situation: Kiran and Anjali have developed a strong mentor-mentee relationship, but they still encounter cultural differences from time to time

    Problem: Anjali is hesitant to discuss certain topics with Kiran, as she fears he may not understand her cultural perspective. Kiran is also afraid of unintentionally offending Anjali with his advice.

    Pain: Anjali feels like she can't be her true self around Kiran, while Kiran feels like he is walking on eggshells around Anjali.

    Solution: Kiran and Anjali openly discuss their cultural differences and agree to approach their conversations with sensitivity and respect. Kiran also takes the time to educate himself on the cultural nuances of Anjali's background. Anjali appreciates Kiran's effort and feels more comfortable discussing sensitive topics with him. They continue to build a strong mentor-mentee.

5. Trust Issue

Situation: Anjali has been working with Kiran for a few months now, but she starts to feel like Kiran is not fully invested in her success.

Problem: Anjali notices that Kiran cancels meetings at the last minute or doesn't follow up on their action items. She starts to wonder if Kiran is losing interest in their mentor-mentee relationship.

Pain: Anjali feels like she can't rely on Kiran and starts to doubt his intentions. She starts to consider finding a new mentor.

Solution: Anjali brings up her concerns with Kiran and asks for his honest feedback. Kiran explains that he has been dealing with some personal issues and apologized for not being fully present in their meetings. He assures Anjali that he is still committed to helping her achieve her goals and agrees to be more transparent about any personal issues that impact their meetings in the future.

6. **Honesty and Transparency**

Situation: Anjali is facing a difficult business challenge and seeking advice from Kiran.

Problem: Kiran is hesitant to provide direct feedback and tries to sugar-coat his advice, fearing that he may hurt Anjali's feelings.

Pain: Anjali feels like Kiran is not being honest with her and starts to question if he is the right mentor for her.

Solution: Kiran realizes that his approach is not helpful to Anjali and decides to be more transparent in his feedback. He explains that his intention is not to hurt her feelings but to help her improve her business. Anjali appreciates Kiran's honesty and becomes more receptive to his advice.

7. **Building Rapport**

Situation: Anjali and Kiran have been working together for a while, but they still struggle to connect on a personal level.

Problem: Anjali feels like Kiran is too focused on the business aspect of their mentor-mentee relationship and doesn't show much interest in her personal life.

Pain: Anjali feels like she can't fully trust Kiran and start to wonder if he really cares about her as a person.

Solution: Kiran suggests that they spend some time outside of their regular meetings to get to know each other better. They plan to attend a business conference together and use the opportunity to bond over shared interests. Anjali appreciates Kiran's effort to build a personal connection and feels more comfortable sharing her personal challenges with him.

8. **Active Listening**

    Situation: Anjali is struggling with a business decision and needs Kiran's guidance

    Problem: Kiran is quick to provide advice without fully understanding Anjali's perspective and needs

    Pain: Anjali feels like Kiran is not listening to her and providing generic advice that is not helpful in her situation

    Solutions: Kiran realizes that he needs to practice active listening to better understand Anjali's challenges and goals. He starts to ask more questions and clarify his understanding of Anjali's situation before providing advice. Anjali appreciates Kiran's effort to understand her better and finds his advice more relevant to her needs.

9. Goal Setting

Situation: Anjali and Kiran are discussing the next steps for her business and setting goals for the upcoming year.

Problem: Anjali has a lot of ideas for her business, but she is not sure which ones to prioritize. Kiran wants to help her set realistic goals, but he is not sure what Anjali's pain points are and what her ultimate vision is for the business.

Pain: Anjali feels overwhelmed and unsure of which direction to take her business. Kiran feels like he is not able to provide the guidance Anjali needs to set achievable goals.

Solution: Kiran starts by asking Anjali about her long-term vision for the business. He then helps her identify the pain points she is currently facing and the goals she wants to achieve in the next year. Kiran helps Anjali break down her goals into smaller, achievable tasks and encourages her to prioritize the ones that will have the greatest impact on her business.

10. Setting Expectations

Situation: Anjali has started implementing the goals she and Kiran set, but she is struggling to make progress as quickly as she had hoped.

Problem: Anjali is not sure if she is meeting Kiran's expectations and feels like she is falling short of her own expectations as well. Kiran, on the other hand, is not sure how to provide feedback without discouraging Anjali.

Pain: Anjali feels like she is not making progress and is starting to doubt her abilities. Kiran feels like he is not able to provide the support Anjali needs to achieve her goals.

Solution: Kiran and Anjali have an open and honest conversation about their expectations for the mentorship. Kiran emphasizes that his role is to guide Anjali, not to set strict expectations. He encourages Anjali to set realistic goals for herself and to focus on making progress, rather than achieving perfection. Anjali feels more empowered to take ownership of her progress and feels more confident in her abilities as an entrepreneur.

11. Technical Challenges

Situation: Anjali is facing technical challenges with her website, which is impacting her abilities to attract and retain customers.

Problem: Anjali is not technically proficient and is not sure how to fix the issue with her website. Kiran is also not familiar with the technical aspects of running a website and is not sure how to help Anjali.

Pain: Anjali feels like she is losing potential customers and revenue due to technical issues with her website. Kiran feels like he is not able to provide the technical support Anjali needs to overcome this challenge.

Solution: Kiran and Anjali research and identify potential solutions to the technical issues with her website. Kiran also recommends that Anjali hire a freelancer or consultant to help her with the technical aspects of her business. Anjali is able to fix the issue with her website and sees an immediate improvement in her customer retention rates.

12. Technical Support

Situation: Anjali has hired a freelancer to help her with the technical aspects of her business, but she is struggling to communicate her need and expectations to the freelancer.

Problem: Anjali is not technically proficient and is not sure how to communicate her needs to the freelancer. The freelancer, on the other hand, is not familiar with Anjali's business and is not sure with her expectations are.

Pain: Anjali feels like the freelancer is not meeting her expectations and is not delivering the results she needs to improve her business. The freelancer feels like he is not able to provide the support Anjali needs without clearer direction from her.

Solution: Kiran recommends that Anjali schedule regular check-ins with the freelancer to provide feedback and clarify any misunderstandings. He also advises Anjali to clearly outline her business goals and expectations for the freelancer to follow. By building trust and rapport with the freelancer through open communication, Anjali can improve the technical support she receives and ultimately, the success of her business.

13. Lack of Direction

Situation: Anjali is facing difficulty in creating a clear direction for her business, and she seeks guidance from Kiran, her mentor.

Problem: Anjali struggles to articulate her vision and goals for the future, and this lack of direction is affecting her ability to make progress.

Pain: Anjali feels overwhelmed and unsure of the steps she needs to take to achieve her goals. She is concerned that she is wasting her time and resources without a clear direction.

Solution: Kiran helps Anjali break down her long-term goals into smaller, more manageable objectives. They work together to identify the specific actions that Anjali needs to take to achieve these objectives. Kiran encourages Anjali to create a detailed action plan that outlines the steps she needs to take to achieve her goals.

14. Competing Priorities

Situation: Anjali is struggling to balance the demands of her personal life with the demands of her business. She seeks guidance from Kiran on how to manage competing priorities.

Problem: Anjali feels overwhelmed by the number of tasks on her to-do list and the limited amount of time she has available to complete them. She is struggling to prioritize her tasks effectively.

Pain: Anjali feels like she is constantly playing catch-up and is unable to make progress on her goals. She is worried that her business will suffer if she is unable to manage her time more effectively.

Solution: Kiran helps Anjali create a system for prioritizing her tasks based on their level of urgency and importance. They also discuss the importance of setting boundaries and learning to say no to tasks that do not align with her priorities.

15. Time Management

Situation Anjali is struggling to manage her time effectively and is not making progress on her business goals as a result.

Problem: Anjali feels like she is constantly busy but not productive. She is not making effective use of her time and is struggling to prioritize her tasks.

Pain: Anjali feels overwhelmed and stressed. She is worried that she will not be able to achieve her goals if she cannot manage her time more effectively.

Solution: Kiran helps Anjali develop a time management system that works for her. They work together to identify the tasks that are most important and create a schedule that allows Anjali to focus on these tasks. Kiran also encourages Anjali to take breaks and prioritize self-care to avoid burnout.

16. Personal Development

Situation: Anjali is interested in personal development and wants to improve her leadership skills. She seeks guidance from Kiran on how to achieve this.

Problem: Anjali is not sure where to start when it comes to personal development. She is unsure which skills she needs to focus on and how to develop them.

Pain: Anjali feels like she is not reaching her full potential as a leader. She is worried that her lack of development will hold her back in her career.

Solution: Kiran helps Anjali identify the specific leadership skills she needs to develop and recommends resources that can help her develop these skills. He also encourages Anjali to seek feedback from her employees and colleagues to identify areas for improvement. Kiran and Anjali work together to create a development plan that allows Anjali to achieve her goals.

17. Career Advancement

Situation: Anjali's business has grown significantly with Kiran's guidance, and she is now managing a team of employees. However, she is struggling to lead effectively and is unsure how to develop her leadership skills.

Problem: Anjali feels overwhelmed and unsure of how to manage her team. She is also unsure of how to communicate her vision effectively and motivate her team.

Pain: Anjali is concerned about the impact her lack of leadership skills may have on her team's morale and productivity. She is also worried that she may not be able to achieve her business goals without effective leadership.

Solution: Kiran suggests that Anjali invests time in developing her leadership skills. He recommends reading books on leadership and attending leadership training programs. He also suggests that Anjali take the time to understand her team's strengths and weaknesses and develop a clear vision for her business so that she can communicate effectively to her team.

18. Performance Reviews

Situation: Anjali's business has continued to grow, and she has recently hired new employees. However, she is struggling to provide effective feedback to her team.

Problem: Anjali finds it challenging to give honest feedback to her employees, fearing that it may hurt their feelings or negatively impact their motivation.

Pain: Anjali is worried that her team may not be performing to the best of their abilities without proper feedback. She is also concerned that her reluctance to give honest feedback may negatively impact her relationships with her employees.

Solution: Kiran helps Anjali understand the importance of regular performance reviews with her employees and how to structure them effectively. He also provides guidance on how to give feed in a constructive and supportive manner. He recommends that she sets clear expectations and goals for her team and provides constructive feedback on their performance. Kiran also advises Anjali to make time for one-on-one meetings with each team member to discuss their individual goals and career development. Kiran encourages Anjali to focus on the positive aspects of her team members' performance and provide actionable steps for improvement.

19. Work-Life Balance

Situation: Anjali is struggling to balance her work responsibilities with her personal life and feels like she is constantly on the go.

Problem: Anjali is struggling to find a balance between her work and personal life. She feels like she needs to be constantly available to her team and customers, and as a result, she has neglected her personal life.

Pain: Anjali is not enjoying the success she has achieved because she feels like she is always working.

She is also worried that she may be neglecting her health and important personal relationships.

Solution: Kiran advises Anjali to prioritize her health and well-being. He suggests that she takes regular breaks and schedule time for herself, such as going to the gym or spending time with family and friends. He also recommends that Anjali delegates some of her responsibilities to her team members to reduce her workload.

20. Transitioning Out

Situation: Anjali has achieved significant success in her business and is ready to take on new challenges. She is considering transitioning out of her current role and exploring new opportunities.

Problem: Anjali is unsure of how to transition out of her mentor-mentee relationship with Kiran. She is grateful for his guidance and doesn't want to damage their relationship.

Pain: Anjali feels conflicted about ending the mentorship with Kiran because she values his guidance and support. She is also worried about losing a valuable mentor and friend.

Solution: Kiran assures Anjali that their mentor-mentee relationship can continue informally. He suggests that they schedule regular catch-up meetings to discuss her progress and any challenges she may be facing. Kiran also encourages Anjali to continue developing her skills by attending networking events and joining relevant industry associations.

21. Continuing the Relationship

Situation: Anjali has successfully transitioned out of her mentorship with Kiran but wants to maintain their relationship.

Problem: Anjali is not sure how to continue their relationship now that the mentorship has ended. She is also worried that Kiran may not have time for her now that she is no longer his mentee.

Pain: Anjali feels conflicted about ending the mentorship with Ravi because she values his guidance and support. She is also worried about losing a valuable mentor and friend.

Solution: Kiran and Anjali discuss ways to maintain the relationship and Kiran assures Anjali that he is still there for her whenever she needs him. They schedule regular catch-up meetings and agree to stay in touch via phone and email. Kiran also suggests that Anjali looks for opportunities to give back, such as by mentoring others or volunteering in the community.

This will help her continue to grow and develop her skills while also giving her a sense of purpose.

With Kiran's support, Anjali is able to continue building on the foundation of trust and rapport that they established during their mentorship. She feels empowered to take on new challenges and pursue her goals, knowing that she has a mentor and friend who believes in her. In the end, the mentor-mentee relationship has grown into a lifelong friendship, based on mutual respect, trust, and a shared commitment to success.

KEY TAKEAWAYS

CHAPTER 2: FINDING THE RIGHT MENTOR: HOW TO IDENTIFY AND CONNECT WITH THE BEST MENTOR FOR YOU

Knowing What You Want: Defining Your Goals And Objectives

1. Be open to different mentorship formats: Mentors can provide guidance through various formats, such as one-on-one meetings, group sessions, or online interactions. Consider what format would work best for you and be open to different approaches.

2. Seek mentors with diverse perspectives: Look for mentors who can offer diverse perspectives and experiences. Different viewpoints can enrich your learning journey and provide a well-rounded understanding of your goals.

3. Set clear expectations: When approaching a potential mentor, be clear about your expectations for the mentorship. Communicate the level of commitment, frequency of interactions, and specific areas you want to focus on.

4. Be receptive to feedback and learning: A good mentor will provide constructive feedback and challenge you to grow. Be open to receiving feedback and willing to learn from your mentor's insights and experiences.

5. Take initiative and be proactive: Mentorship is a two-way street. Take initiative in the mentorship relationship by asking questions, seeking advice, and actively participating in discussions and activities.

6. Be patient and persistent: Finding the right mentor may take time. Be patient in your search and don't be discouraged by setbacks. Keep pursuing opportunities and stay persistent in achieving your goals.

By defining your goals and objectives, you can align your mentorship journey with your aspirations and maximize the benefits of having a mentor who can guide and support you on your path to success.

Building Your Network: How To Meet Potential Mentors

1. Be Active on LinkedIn: Engage with industry professionals, join relevant groups, and participate in discussions to connect with potential mentors.

2. Attend Industry Conferences and Events: Actively participate in conferences and events in your field to meet experts and leaders who could become your mentors.

3. Leverage Alumni Networks: Connect with alumni from your educational institution or previous workplaces, as they might be willing to mentor and support you.

4. Seek Referrals: Ask friends, colleagues, or acquaintances if they know someone who could be a potential mentor for you.

5. Utilize Professional Associations: Join industry-specific associations and engage with their members to find mentors with relevant expertise.

6. Be Genuine and Authentic: When reaching out to potential mentors, be genuine and authentic in your approach. Show your passion for learning and growth.

7. Be Open to Different Perspectives: Be willing to connect with mentors from diverse backgrounds and industries to gain broader insights.

8. Offer Value to Your Mentor: Show how you can contribute to your mentor's growth or projects, creating a mutually beneficial relationship.

9. Stay Persistent and Respectful: Building a mentorship relationship may take time, so be persistent in your efforts, but also respectful of the mentor's time and commitments.

Remember, mentorship is a valuable opportunity to learn and grow, and by actively seeking out and engaging with potential mentors, you can significantly enhance your personal and professional development.

Qualities To Look For In A Mentor: What To Consider When Choosing A Mentor

1. Seek a mentor who is open and honest, as their willingness to share experiences and mistakes can help you learn and grow.

2. Look for a mentor who is flexible and adaptable, as they can help you navigate a rapidly changing industry or profession.

3. Find a mentor who exudes passion and enthusiasm, as their energy can inspire and motivate you to pursue your goals.

4. Choose a mentor with strong intuition and emotional intelligence, as they can help you build strong relationships and navigate complex situations.

5. Look for a mentor with strategic thinking and vision, as their guidance can help you develop long-term planning skills.

Key Takeaways on the Worst Qualities of a Mentor:

1. Avoid mentors who are self-centered, as they may prioritize their own success over your growth and development.

2. Stay away from mentors lacking empathy, as they may not understand or support your perspective and feelings.

3. Avoid mentors who are dishonest or manipulative, as their actions can break trust and put your career at risk.

4. Be cautious of mentors who are overly controlling, as they may hinder your growth by preventing you from taking risks and making decisions.

5. Look for mentors with relevant experience in your field or industry, as those lacking experience may not provide effective guidance and support.

By seeking mentors with positive qualities and avoiding those with negative traits, you can maximize the impact of mentorship on your personal and professional growth.

Approaching Your Mentor: How To Initiate The Mentorship Relationship

1. Be persistent and demonstrate enthusiasm: Don't be discouraged if a potential mentor is initially hesitant. Show your dedication and passion for the field, and your persistence may lead to a mentorship opportunity.

2. Be specific and clear in your request: Clearly express your interest in a mentorship relationship and explain why you believe they would be a valuable mentor for you.

3. Utilize different channels: Reach out through networking events, social media, alumni networks, or professional organizations to find potential mentors who align with your goals.

4. Follow up and stay engaged: Keep the communication lines open and show your progress and commitment to the mentorship relationship. This will demonstrate your eagerness to learn and grow.

5. Seek mentors in relevant fields: Look for mentors who have experience and expertise in the specific areas you want to develop. This will ensure their guidance is relevant and beneficial to your goals.

6. Take a direct approach if necessary: If you have identified a mentor, you believe is perfect for you, don't hesitate to approach them directly with a well-thought-out request for mentorship.

7. Build a strong foundation: Establish trust and respect in the mentor-mentee relationship by actively listening, seeking feedback, and demonstrating gratitude for their support.

Building Trust and Rapport: How To Develop A Strong Mentor-Mentee Relationship

1. Communicate openly and honestly.
2. Be sensitive to cultural differences.
3. Practice active listening.

4. Provide honest feedback.
5. Set achievable goals together.
6. Prioritize work-life balance.
7. Stay connected even after the formal mentorship ends.

CHAPTER 3: LEARNING FROM MENTORS: STRATEGIES AND TECHNIQUES FOR MAXIMIZING THE BENEFITS OF MENTORSHIP

"Tell me and I forget, teach me and I may remember, invoice me and I learn"
– Benjamin Franklin

Mentorship is a powerful tool that can propel individuals towards success and personal growth. In this chapter, we will explore strategies and techniques that will help you maximize the benefits of mentorship. By learning knowledge from mentors, you can gain valuable insights, expand your knowledge, and accelerate your progress towards your goals. This chapter will provide you with practical advice on how to establish effective mentorship relationships, leverage the wisdom of mentors, and make the most of this transformative experience.

Building Meaningful Mentorship Relationships:

To make the most of mentorship, it is crucial to establish meaningful relationships with your mentors. One effective strategy is to seek out mentors who align with your goals and values. By choosing mentors who have achieved what you aspire to, you can tap into their expertise and learn from their experiences. Additionally, it is essential to demonstrate your commitment and dedication to the mentorship relationship. Show genuine interest in their guidance and actively seek their feedback. Regularly scheduled meetings and open communication channels can foster a strong mentor-mentee bond.

Leveraging the Wisdom of Mentors:

Once you have established a mentorship relationship, it is essential to leverage the wisdom and knowledge of your mentors effectively. Begin by setting clear goals and expectations for the mentorship journey. Communicate your objectives and seek guidance on how to achieve them. Actively listen to your mentors' advice and take their feedback seriously. Remember that their insights are based on years of experience and can help you navigate challenges and avoid common pitfalls. Be open to constructive criticism and embrace a growth mindset, as this will enable you to make the most of the mentor's guidance.

Making the Most of the Transformative Experience:

To maximize the benefits of mentorship, it is important to actively engage in the learning process. Take the initiative to identify areas where you need improvement and seek guidance from your mentor on how to develop those skills. Actively participate in workshops, seminars, and networking opportunities recommended by your mentor. Embrace a proactive approach to learning and continuously seek new challenges to broaden your horizons. Regularly reflect on your progress, celebrate your achievements, and identify areas for further growth. By actively applying the knowledge and insights gained from your mentor, you can truly make the most of this transformative experience.

Mentorship is a powerful and effective tool for personal and professional development. By implementing the strategies and techniques outlined in this chapter, you can maximize the benefits of mentorship. Remember to build meaningful relationships with your mentors, leverage their wisdom effectively, and actively engage in the learning process. Mentorship offers a unique opportunity to gain insights, expand your skills, and accelerate your journey towards success. Embrace this transformative experience and let the guidance of mentors propel you towards achieving your goals and unlocking your full potential.

- **Gain Perspective and Insight:** Mentors can provide valuable perspective and insight into your personal and professional life. They can help you see things from a different point of view and provide guidance on how to navigate challenges.
- **Develop New Skills:** Mentors can help you develop new skills and knowledge that can be applied to your

personal and professional life. They can provide guidance and support as you work towards improving your skills and reaching your goals.

- **Build Confidence:** A mentor can provide encouragement and support, helping you build confidence in yourself and your abilities. This can help you overcome self-doubt and achieve success in your personal and professional life.

- **Expand Your Network:** Mentors can introduce you to new people and expand your professional network. This can lead to new opportunities, collaborations, and career advancements.

- **Gain Career Insights:** Mentors can provide insights into your industry or profession, helping you better understand career paths and opportunities. They can also share their own experiences and offer guidance on how to advance their career.

Maximize the Benefits of Mentorship:

- **Be Open and Willing to Learn:** Approach mentorship with an open mind and a willingness to learn. Be receptive to feedback and willing to make changes to improve yourself and your career.

- **Set Clear Goals:** Identify your goals and objectives before seeking a mentor. This will help you find a mentor, who can provide the guidance and support you need to achieve your goals.

- **Build a Strong Relationship:** Develop a strong relationship with your mentor based on trust and respect. Be open and honest with your mentor, and actively seek their guidance and feedback.

- **Take Action:** Applying the advice and feedback from your mentor is critical to achieving your goals. Use what you learn to progress in your personal and professional life and be willing to take risks and try new things. Taking action is the key to making real progress and achieving your goals.

- **Reflect on Your Progress:** Regularly assess your progress and reflect on what you've learned. Celebrate your successes, identify areas for improvement, and adjust your goals as needed. Reflecting on your progress and your mentor's guidance can help you stay motivated and focused.

Setting Expectations: Establishing Clear Goals and Objectives

1. Define the Goals and Objectives: Clearly define what you want to achieve and why it is important. This will provide a clear understanding of what needs to be done and why it matters. This will help you to communicate clearly with your mentor.

2. Discuss your expectations: Discuss your expectations with your mentor early on. This includes discussing how often you will meet, what you hope to gain from the mentorship, and what you expect from your mentor.

3. Break down the Goals and Objectives: Break down the goals and objectives into smaller, more manageable steps. This will help you to avoid feeling overwhelmed and provide a clear path forward.

4. Make the Goals and Objectives SMART: Make the goals and objectives Specific, Measurable, Achievable, Relevant, and Time-bound. This will make them more tangible and easier to measure progress.

5. Regularly check in: Regularly check in with your mentor to track progress, discuss any challenges, and make adjustments as needed. This will help you to stay on track and make progress towards your goals.

6. Involve Stakeholders: Involve stake-holders, including team members, partners, and customers, in the process of setting expectations and establishing goals and objectives. This will help to ensure that everyone is aligned and working towards the same outcomes.

7. Communicate Clearly: Communication expectations clearly and frequently including progress updates and changes to the goals or objectives. This will ensure that everyone is on the same page and working towards the same outcomes.

8. Hold yourself accountable: Hold yourself accountable for achieving the goals and objectives. This will ensure that everyone is on the same page and working towards the same outcomes. This involves

setting deadlines and milestones, tracking progress, and taking responsibility for any setbacks or failures.

9. Adjust as needed: Be willing to adjust the goals and objectives as needed based on new information or changing circum-stances. This will ensure that the goals remain relevant and achievable.

Communication Skills: How to Communicate Effectively with Your Mentor

To achieve your desired goals, effective communication with your mentor is paramount. The mentor-mentee relationship relies heavily on communication and a lack thereof will hinder progress and potentially impede success. In order to maximize the benefits of mentorship, clear communication is vital to ensure that you both are on the same page and are able to overcome roadblocks and challenges. By communicating effectively with your mentor, you can gain a fresh perspective and potentially achieve your goals more efficiently than anticipated. These steps provide guidance on how to effectively communicate with your mentor in the face of challenges and setbacks, allowing you to achieve your goals with greater ease and success.

- Be open to feedback: One of the most valuable aspects of mentorship is receiving feedback from someone who has more experience than you. Be open to feedback, even if it is difficult to hear, and use it to improve your skills and performance.

- Listen actively: When communicating with your mentor, be sure to listen actively. This means fully engaging in the conversation, asking questions for clarification, and summarizing what you've heard to ensure that your understanding is correct.

- Use your mentor as a sounding board: Use your mentor as a sounding board to help you think through your challenges and identify potential solutions. They can provide a fresh perspective and help you see things from a different angle.

- Learn from failure: Don't be afraid to fail. Failure can be a valuable learning opportunity and can help you grow and develop as a leader. Use your mentor's guidance and support to help you learn from failure and move forward.

- Analyze the situation: Take a step back and analyze the situation objectively. Identify what went wrong and why, and what you could have done differently.

- Focus on solutions: Instead of dwelling on the challenges, focus on finding solutions. Brainstorm with your mentor on ways to overcome the challenges and take action on the most feasible and effective solutions.

- Learn from their experience: Mentors have valuable experiences that can help you navigate challenges. Ask your mentor about their own experience with similar challenges and how they overcame them. This can provide you with valuable insights and guidance.

- Embrace a growth mindset: Adopt a growth mindset and view challenges as opportunities for growth and development. Be open to learning from your mistakes and failures and use them as learning experiences.

- Practice resilience: Resilience is the ability to bounce back from setbacks. Practice resilience by staying positive, focusing on your strengths, and staying committed to your goals.

- Be persistent: Remember that overcoming challenges takes time and persistence. Don't give up if progress is slow and stay committed to achieving your goals.

- Be proactive: Don't wait for your mentor to take the lead. Be proactive in setting up meetings, asking for feedback, and following up on action items.

- Follow through on action items: If your mentor gives you specific action items to work on, be sure to follow through on them. This will show your mentor that you are committed to your development and value their input.

- Stay accountable: Set clear goals and deadlines for addressing the challenges and share them with your mentor. Hold yourself accountable for making progress and provide updates to your mentor on your progress.

- Be respectful of your mentor's time: Your mentor is likely a busy person, so it's important to be respectful of their time. Come to meetings prepared and ready to discuss specific topics or issues.

- Celebrate Successes: When you do overcome challenges and achieve your goals, celebrate your success with your mentor. This will help reinforce positive behaviours and keep you to continue learning and growing.

Active Listening: How to Listen to Your Mentor and Learn from Their Advice

Active listening is a critical skill that can help you learn and grow from your mentor's advice. Emma's experience can demonstrate the significant impact that active listening can have on your life.

The Beginning of the Journey

Date: March 3, 2007

City: San Francisco, USA

Emma was a recent college graduate eager to start her career in marketing. She was ambitious and had big dreams, but she knew that she needed guidance to achieve her goals. She reached out to a successful marketing executive, Jennifer, who had years of experience in the industry and asked her to be her mentor.

The First Lesson

Date: April 10, 2007

City: San Francisco, USA

Jennifer agreed to be Emma's mentor and they started meeting regularly. During their first meeting, Jennifer asked Emma what her career goals were. Emma replied that she wanted to become a marketing director by the age of 30. Jennifer listened carefully and then asked Emma about her experience in the industry so far. Emma excitedly talked about her internships and the projects she had worked on. Jennifer then gave her some advice: "Emma, to become a marketing director, you need to have a deep understanding of your industry and your customers. You need to be willing to learn and grow constantly. Don't be in a rush to climb the ladder. Focus on building your skills and knowledge first."

The First Test

Date: September 15, 2008

City: San Francisco, USA

A year later, Emma was offered a job at a leading marketing firm. She was thrilled and eager to prove herself. She worked hard, put in long hours, and took on every project that came her way. But as time passed, Emma started to feel burnt out. She was struggling to balance her workload and her personal life. She reached out to Jennifer for advice. Jennifer listened to Emma's concerns and reminded her of the advice she had given her a year ago. She encouraged Emma to take a step back, focus on her well-being and prioritize her goals.

The Major Challenge

Date: May 12, 2010

City: San Francisco, USA

Two years later, Emma was faced with a major challenge. The marketing firm she worked for was acquired by a larger company, and her job was on the line. Emma was devastated. She had worked so hard to get where she was, and not it all seemed to be slipping away. She reached out to Jennifer for advice. Jennifer listened to Emma's concerns and reminded her of her own experiences. She shared how she had faced similar challenges in her career and how she had learned to pivot and adapt. She encouraged Emma to stay positive, keep learning, and be open to new opportunities.

The Outcome

Date: June 15, 2011

City: San Francisco, USA

Over the years, Emma continued to meet with Jennifer regularly. She learned to listen carefully to her mentor's advice and apply it to her own life and career. She focused on building her skills and knowledge, taking on new challenges, and prioritizing her goals. As a result, Emma was able to achieve her dream of becoming a marketing director by the age of 30. She also found a balance between her personal and professional life and built a fulfilling career in the marketing industry.

Emma's success can be attributed to her application of Jennifer's advice, which emphasizes the importance of active listening and learning from mentorship. This lesson is applicable to everyone, as actively listening to our mentors can lead to valuable insights and guidance that can help us achieve our goals.

Be Open-Minded: Emma remained receptive to every perspective or point of view that Jennifer shared with her, which enabled her to broaden her knowledge and enhance her skills.

Pay attention: When your mentor speaks, give them full attention. Avoid getting distracted by your phone, computer, or anything else. Stay present in the moment and focus on what your mentor is saying. Emma remained fully focused on her mentor Jennifer during their meetings, never getting distracted by her phone, computer, or anything else.

Ask questions: Clarify anything you don't understand by asking questions. This not only shows that you're actively listening but also helps you better understand the advice you're receiving. It is always recommended to summarize what you have heard to confirm that you have comprehended it accurately. Emma was proactive in asking questions to ensure that all her doubts were cleared.

Take notes: Taking notes is a great way to ensure that you don't forget any important points your mentor makes. You can refer back to your notes later to refresh your memory. Emma diligently took notes and incorporated every crucial point that her mentor Jennifer imparted into her life.

Seek out challenges: Be willing to take on new challenges and step out of your comfort zone. Your mentor can provide guidance and support as you tackle new projects or tasks, and this can help you to grow and develop new skills. Emma actively pursued new challenges and consistently strived towards achieving her goals, refusing to remain within her comfort zone.

Take action: The most effective way to learn from your mentor is to take action on their advice. Put their suggestions into practice and see how they work for you. This will not only show your mentor that you are serious about learning from them, but it will also help you to grow and improve. Emma diligently acted upon every task assigned to her by her mentor Jennifer, which enabled her to achieve her objectives promptly and precisely within the specified timeframe.

Taking Actions: How to Apply What You Learn from Your Mentor:

Taking action is crucial when it comes to maximizing the benefits of mentorship. One of the key strategies is to set clear goals and objectives for your mentorship. These should be specific, measurable, achievable, relevant, and time-bound (SMART) goals that align with your overall career aspirations. To apply the strategies and techniques learned from your mentor, it's important to take a proactive approach to your personal and professional development. Start by reflecting on what you've learned from your mentor and identifying specific areas where you can apply these insights in your daily work. This could involve adopting new habits, developing new skills, or refining your existing processes to be more efficient and effective. It's also important to prioritize your action items based on their potential impact and feasibility and to track your progress over time to stay on track and make adjustments as needed. Additionally, consider seeking out additional resources or support to help you implement your mentor's advice, such as training programs, workshops, or networking events. Finally, don't be afraid to experiment and take risks in your pursuit of personal and professional growth. By taking action on what you learn from your mentor, you can develop new competencies, achieve your goals, and position yourself for long-term success.

Increased Productivity:

- Identify your most important tasks and prioritize them accordingly.
- Use time-blocking to schedule dedicated periods for specific tasks.
- Minimize distraction by turning off notifications or working in a quiet space.
- Take regular breaks to avoid burnout and maintain focus.
- Use productivity tools such as project management software, task list, or timers to help stay on track.

Enhanced Performance:

- Practice self-reflection to identify areas for improvement.
- Work with your mentor to develop a plan for improving your skills or knowledge in these areas.
- Focus on continuous learning and development to stay up to date with industry trends.
- Ask for input from your mentor to assist you in recognizing areas where you can enhance your skills.
- Take on challenging assignments to push yourself beyond your comfort zone.

Expanded Knowledge and Skills:

- Attend training programs, workshops, or industry events to learn new skills.

- Explore possibilities to collaborate with colleagues from various departments to broaden your understanding of different facets of the organization.
- Engage in self-directed learning by reading books, taking online courses, or watching webinars.
- Practice your skills through hands-on experience by seeking your mentor's help to provide guidance and feedback.
- Share your knowledge and skills with others through mentorship or teaching opportunities.

Better decision-making:

- Work with your mentor to identify specific decision-making challenges.
- Discuss and analyze past decisions to identify areas for improvement.
- Look for new perspectives and information to influence your process of decision-making.
- Consider potential risks and benefits before making a decision.
- Test your assumptions through small-scale experiments or prototypes.
- Learn from your mistakes and use them to influence future decisions.

Career Growth:

- Set clear career goals and develop a plan to achieve them.

- Look for opportunities to work on high-impact projects or initiatives.
- Build a professional network by attending networking events or connecting with industry professionals.
- Take on leadership roles or responsibilities to develop your management skills.
- Request your mentor to provide guidance and support to your career development.

Improved Self-Confidence:

- Work with your mentor to identify areas where you feel less confident.
- Develop a plan for building your self-confidence, such as practicing your public speaking or seeking out feedback on your work.
- Practice self-care by getting enough rest, exercise and healthy food to reduce stress and improve your overall well-being.
- Celebrate your successes and reflect on what you've learned from your failures.
- Surround yourself with positive and supportive people who lift you up.
- Take support from your mentor when you need it.

Overcoming Challenges: How to Deal with Common Obstacles in Mentorship

Overcoming challenges is an essential part of any mentorship relationship, and common obstacles can arise in various forms. One effective approach to deal with these challenges is to establish clear communication channels between the mentor and the mentee. Another way is to set realistic goals and expectations, which can help in building a more productive and constructive relationship. However, these strategies may not always work, and mentorship relationships can be tested by unforeseen challenges. A real incident that highlights the importance of perseverance and adaptability in mentorship can provide a clear understanding of how to overcome challenges and achieve success in the relationship.

Mentor: Jane Smith

Mentee: Sarah Jones

Location: New York City

Year: 2015

Overthinking: Learning to Trust the Process

Situation: Sarah Jones was a mentee of Jane Smith, a successful entrepreneur who was guiding Sarah to start her own business. However, Sarah was constantly overthinking every decision she made and was paralyzed by the fear of failure.

Pain: Sarah felt stuck and frustrated with herself for not being able to move forward with her business plans.

Problem: Sarah's overthinking was preventing her from taking any action towards her goals, and it was causing her to doubt herself and her abilities.

Solution: Jane encouraged Sarah to trust the process and to focus on taking small steps towards her goals. She suggested Sarah break down her larger goals into smaller achievable ones and celebrate each milestone along the way. Additionally, Jane recommended that Sarah practice mindfulness techniques to help calm her overactive mind and learn to recognize and let go of her negative thought patterns.

Lack of Commitment: Finding Purpose and Passion

Situation: Sarah had a lot of ideas and goals for her business, but she was having trouble committing to any of them. She felt overwhelmed and uncertain about what direction to take.

Pain: Sarah felt unfulfilled and frustrated with herself for not being able to commit to anything.

Problem: Sarah's lack of commitment was preventing her from making progress towards her goals and was causing her to feel stuck.

Solution: Jane worked with Sarah to identify her passions and to help her define her purpose. They discussed the reasons behind her goals and what motivated her. This process helped Sarah to clarify her goals and prioritize them. Additionally, Jane suggested that Sarah create a detailed plan with actionable steps to achieve her goals and hold herself accountable by regularly reviewing her progress.

Communication Breakdown: Active Listening and Open Communication

Situation: Sarah and Jane had different communication styles, which sometimes caused misunderstandings and tension between them.

Pain: Sarah felt frustrated with the lack of clear communication and was unsure of how to move forward.

Problem: The breakdown in communication was hindering their ability to work effectively together and could potentially harm their mentor-mentee relationship.

Solution: Jane and Sarah had an open and honest conversation about their communication styles and how they could improve. They worked on active listening techniques and using clear and concise language when communicating. Additionally, they set regular check-ins to discuss any potential issues and to ensure they were on the same page.

Conflicting Schedules: Flexibility and Time Management

Situation: Sarah and Jane had busy schedules, which sometimes made it difficult to find time to meet.

Pain: Sarah felt like her progress was being hindered by conflicting schedules and felt frustrated that she couldn't make more progress.

Problem: The conflicting schedules were making it difficult for Sarah to get the guidance and support she needed from Jane.

Solution: Jane and Sarah worked together to find ways to make their schedules work. They scheduled regular meetings in advance and were flexible when unexpected conflicts arose. Additionally, they explored alternative methods of communication, such as email and phone calls, to stay connected and make progress even when they couldn't meet in person.

Different Learning Styles: Tailoring the Approach

Situation: Sarah and Jane had different learning styles, which sometimes made it challenging for Jane to effectively teach and guide Sarah.

Pain: Sarah felt like she wasn't understanding the concepts and was struggling to apply them to her business.

Problem: The different learning styles were hindering Sarah's ability to learn and apply new concepts, which was preventing her from making progress towards her goals.

Solution: Jane worked with Sarah to understand her learning style and tailored her approach to fit Sarah's needs. They used a variety of teaching methods, including visual aids and hands-on learning, to help Sarah understand and retain the information. Additionally, Jane encouraged Sarah to ask questions and provide feedback, so that she could adjust her teaching approach as needed.

Lack of Trust: Building a Strong Relationship

Situation: Sarah had a hard time trusting Jane and was hesitant to share her vulnerabilities and struggles with her.

Pain: Sarah felt isolated and unsupported, which was hindering her ability to make progress towards her goals.

Problem: The lack of trust between Sarah and Jane was preventing them from building a strong mentor-mentee relationship, which was essential for Sarah's success.

Solution: Jane worked to build trust with Sarah by being transparent and open about her own struggles and experiences. She listened to Sarah without judgement and provided honest feedback and support. Additionally, Jane encourages Sarah to take risks and make mistakes, emphasizing that failure was a natural part of the learning process.

Family Challenges: Balancing Priorities

Situation: Sarah had a young child and a spouse, which sometimes made it difficult for her to balance her family responsibilities with her business goals.

Pain: Sarah felt guilty for neglecting her family and frustrated that she couldn't make more progress towards her goals.

Problem: The family challenges were hindering Sarah's ability to focus on her business and were causing her to feel overwhelmed and stressed.

Solution: Jane and Sarah worked together to create a plan that balanced Sarah's family responsibilities with her business goals. They identified times when Sarah could work on her business, such as during her child's nap time, and set realistic expectations for progress. Additionally, Jane encouraged Sarah to involve her family in her business goals, such as by having her spouse take care of their child during important meetings. This helped Sarah to feel supported and motivated to succeed.

Financial Loss and Challenges: Adapting to Change

Situation: Sarah experienced unexpected financial setbacks that threatened her business.

Pain: Sarah felt overwhelmed and stressed, as she was unsure of how to recover from the financial losses.

Problem: The financial losses were hindering Sarah's ability to move forward with her business and were causing her to doubt the viability of her business idea.

Solution: Jane worked with Sarah to adapt her business plan to the new financial reality. They explored new revenue streams and cost-cutting measures, such as outsourcing certain tasks or downsizing the business. Additionally, Jane encouraged Sarah to stay focused on her long-term goals and not to be discouraged by setbacks. They worked together to create a plan to recover from the financial losses and to continue moving forward with the business.

KEY TAKEAWAYS

CHAPTER 3: LEARNING FROM MENTORS: STRATEGIES AND TECHNIQUES FOR MAXIMIZING THE BENEFITS OF MENTORSHIP

Setting Expectations: Establishing Clear Goals and Objectives

- Mentors can provide perspective, insight, skill development, confidence building and network expansion.

- To maximize the benefits of mentorship, be open and willing to learn, set clear goals, build a strong relationship, take action and reflect on your progress.

- When setting expectations define goals and objectives, discuss expectations, break down goals, make them SMART regularly check in, involve stakeholders, communicate clearly, hold yourself accountable and adjust as needed.

Communication Skills: How to Communicate Effectively with Your Mentor

- Effective communication with your mentor is crucial for achieving your goals. You can improve communication by being open to feedback, actively listening, using your mentor as a sounding board, learning from failure, analyzing the situation objectively focusing on solutions, learning from your mentor's experience, adopting a growth mindset, practising resilience, being persistent, being proactive, following through on action items, staying accountable, being respectful of your mentor's time, and celebrating successes. By following these steps, you can strengthen your mentor-mentee relationship, gain valuable insights and guidance, and achieve your goals more efficiently.

Active Listening: How to Listen to Your Mentor and Learn from Their Advice

- Being present and actively listening to your mentor when they share valuable experiences with you is a crucial aspect of developing communication skills. This involves asking questions, clarifying your doubts, and summarizing what you've heard to ensure that you've understood the point correctly. Taking notes is also a helpful way to ensure that you don't miss any important points, and you can refer back to them later to refresh your memory.

Taking Action: How to Apply What You Learn from Your Mentor

- To apply what you learn from your mentor, set clear goals and objectives that align what your career aspirations and take a proactive approach to personal and professional development.

- To increase productivity, prioritize tasks, use time-blocking, minimize distractions, take breaks, and use productivity tools such as project management software or timers.

- To enhance performance practice self-reflection, work with your mentor to develop a plan for improvements, focus on continuous learning and development, and take on challenging assignments.

- To expand knowledge and skills, attend training programs, collaborate with colleagues, engage in self-directed learning, seek hands-on experience, and share knowledge with others.

- To improve decision decision-making, work with your mentor to identify challenges, analyze past decisions, consider new perspectives, test assumptions, and learn from mistakes.

- To achieve career growth, set clear goals, seek high-impact projects or initiatives, build a professional network, take on leadership roles or responsibilities, and seek guidance and support from your mentor.

- To improve self-confidence, identify areas for improvement, develop a plan for building self-confidence, practice self-care, celebrate successes, surround yourself with positive and supportive people, and seek support from your mentor when needed.

Overcoming Challenges: How to Deal with Common Obstacles in Mentorship

- Establish clear communication channels between the mentor and the mentee.

- Set realistic goals and expectations and break larger goals into smaller achievable ones.

- Practice mindfulness techniques to help calm the overactive mind and recognize negative thought patterns.

- Identify passions and define purpose.

- Create a detailed plan with actionable steps to achieve goals.

- Use active listening techniques and clear and concise language when communicating.

- Schedule regular check-ins and be flexible when unexpected conflicts arise.

- Tailor the approach to accommodate different learning styles.

- Build trust with the mentee by being transparent, open and supportive.

- Balance family responsibilities with business goals.

- Adapt to changes in circumstances, such as financial setbacks, and find new solutions to achieve success.

CHAPTER 4: MENTORS IN THE REAL WORLD: CASE STUDIES AND EXAMPLES OF SUCCESSFUL MENTOR-MENTEE RELATIONSHIPS

"A mentor is someone who sees more talent and ability within you, than you see in yourself, and helps bring it out of you"- Bob Proctor

One potential transformational leadership mentor that can be gleaned from studying successful mentor-mentee relationships is the importance of creating a supportive and empowering environment for mentees to learn and grow. Effective mentors empower their mentees by providing opportunities for them to take on challenges and responsibilities, and by fostering an environment of open communication, trust, and respect. Mentors who are able to provide guidance, feedback, and support while also allowing their mentees to take ownership of their development can inspire transformational change and help their mentees reach their full potential. Additionally, mentors who are able to lead by example and model the behaviors and attitudes that they want their mentees to embody can create a culture of transformational leadership within their organizations. By focusing on creating an environment of support and empowerment, mentors can help their mentees develop the skills and mindset needed to become transformational leaders themselves.

Celebrity Mentors: Famous Examples of Mentorship

Here are some examples of successful mentor-mentee relationships with celebrity mentors:

- A. P. J. Abdul Kalam: A. P. J Abdul Kalam was a renowned scientist and former President of India who mentored many young people throughout his life. He believed in the power of education and encouraged young people to pursue their dreams, despite any obstacles they may face. Some of his notable mentees include Srijan Pal Singh, a young social entrepreneur and author, and Shradha Prasad, a young engineer and entrepreneur who co-founded the startup Anthill Ventures.

- Barack Obama: Barack Obama, the former President of the United States, has had many mentors throughout his life. One of his most influential mentors was a man named Frank Marshall Davis, who was a poet and writer. Davis served as a father figure to Obama and has also spoken about the influence of his grandmother, who instilled in him a love of learning and a strong work ethic.

- Satya Nadella: Satya Nadella is the CEO of Microsoft and has spoken publicly about the importance of having mentors throughout his career. One of his most influential mentors was a man named Soma Somasegar helped Nadella to navigate the complexities of the tech industry and provided him with valuable advice and support as he rose through the ranks at Microsoft.

- Sundar Pichai: Sundar Pichai is the CEO of Google and has spoken about the influence of his father, who was an electrical engineer, on his career path. Pichai has also spoken about the importance of having mentors throughout his career and has credited a former boss, Jeff Huber, with helping him to develop his leadership skills and navigate the complex world of technology.

- Rishi Sunak: Rishi Sunak is the Chancellor of the Exchequer in the UK government and has spoken about the influence of his parents on his career. His mother was a pharmacist, and his father was a doctor, and they instilled in him a strong work ethic and a belief in the power of education. Sunak has also credited his former boss, hedge fund manager Patrick Degoce, with helping him to develop his business skills and providing him with valuable mentorship early in his career.

- Parag Agrawal: Parag Agrawal is the CEO of Twitter and has spoken about the importance of mentorship throughout his career. One of his most influential mentors was a man named Chris Manning, who was a professor at Stanford University. Manning helped Agrawal to develop his skills as a data scientist and provided him with valuable advice and guidance as he pursued his career in tech.

- Ratan Tata: Ratan Tata is s renowned Indian businessman and philanthropist who has mentored many young entrepreneurs throughout his career. He has spoken about the importance of giving back and helping others and has provided mentorship and support to many startups and social entrepreneurs in India.

- Narendra Modi: Narendra Modi is the current Prime Minister of India and has spoken about the influence of his mother on his life. Modi's mother worked as a cleaner in a school and instilled in him a strong work ethic and a belief in the power of education. Modi has also credited his spiritual mentor, Swami Vivekananda, with inspiring him to pursue a life of service and leadership.

Mentorship in the Workplace: Success Stories from Corporate America

Mentorship programs have become increasingly popular in corporate America as a way to develop and retain talent. Here are a few success stories from companies that have implemented mentorship programs:

1. PepsiCo: In 2016, PepsiCo launched a mentoring program called "Dare to Lead," which paired high-potential junior employees with senior leaders. The

program provided mentors and mentees with tools to set goals, measure progress, and develop leadership skills. The program was so successful that it expanded to include over 2,500 employees in 2019.

2. Intel: Intel's "Pay it Forward" mentorship program encourages employees to share their knowledge and experience with others. The program is voluntary, and employees can choose to participate as mentors, mentees, or both. The program has been credited with helping employees build their skills, develop their careers, and improve their job satisfaction.

3. Cisco: Cisco's mentorship program, called "The Reverse Mentorship Program," pairs senior executives with younger employees to help them stay up to date with technology trends and emerging market opportunities. The program has been a huge success, with senior leaders learning from their junior counterparts and becoming better equipped to make informed decisions.

4. General Electric: General Electric's mentorship program, called "GE Global Mentoring," pairs employees with mentors from different parts of the world to help them develop a global perspective. The program has helped employees improve their cross-cultural communication skills and gain a better understanding of global business practices.

5. Microsoft: Microsoft's mentorship program, called "Mentor Like a Microsoft," encourages employees to become mentors and help others develop their skills

and achieve their goals. The program provides training and support to both mentors and mentees and has been credited with improving employee retention and engagement.

Overall, mentorship programs can provide significant benefits for both employees and companies. By sharing knowledge and experience, mentors can help their mentees develop their skills and achieve their goals, while companies can benefit from increased employee engagement, retention, and productivity.

Mentorship in Education: How Teachers and Professors Can Impact Students Lives

In 2016, Dr Abhinav Dhotre, an experienced professor at a college in Belgaum, India, took on a new mentee, Vaishnavi, who was a first-year student struggling with her studies. Dr Abhinav's guidance and mentorship helped Vaishnavi not only improve her academic performance but also develop critical life skills.

- **Critical Thinking:** Vaishnavi often struggled with critical thinking and analyzing information, which affected her academic performance. Dr Abhinav guided her on how to analyze information, identify patterns and interpret data. He taught her the importance of critical thinking and how to approach problems from different perspectives. With Dr Abhinav's guidance, Vaishnavi was able to improve her critical thinking abilities and perform better in her exams.

- **Strategic Decision Making:** Vaishnavi found it challenging to make decisions, often getting stuck between two choices. Dr Abhinav taught her how to weigh the pros and cons of different options, evaluate their potential outcomes, and make informed decisions. He also advised her on how to create a long-term plan to help her achieve her goals. Vaishnavi learned how to make strategic decisions and develop a clear path towards success.

- **Problem-Solving:** Vaishnavi often struggled with solving complex problems, which impacted her academic performance. Dr Abhinav taught her how to break down problems into smaller parts, analyze each component, and identify potential solutions. He also encouraged her to think outside the box and come up with creative solutions. With Dr Abhinav's guidance, Vaishnavi was able to improve her problem-solving skills and tackle complex academic problems with confidence.

- **Setting High Expectations:** Dr Abhinav set high expectations for Vaishnavi and held her accountable for meeting those expectations. He encouraged her to set ambitious goals and develop a plan to achieve them. With his guidance, Vaishnavi learned the importance of setting high expectations and striving for excellence in all areas of her life.

- **Motivating Team Members:** Dr Abhinav taught Vaishnavi how to motivate and inspire her team members, which helped her become a more effective team player. He showed her how to identify individual strengths and use them to create a cohesive and motivated team. Vaishnavi learned how to lead by example and encourage her team members to reach their full potential.

- **Teamwork:** Vaishnavi struggled with working effectively in a team, often finding it difficult to communicate her ideas and collaborate with others. Dr Abhinav taught her how to communicate effectively, listen actively, and build trust with her team members. He also showed her how to resolve conflicts and build strong relationships within the team. With Dr Abhinav's guidance, Vaishnavi learned how to work effectively in a team and contribute to the team's success.

- **Communicate Effectively:** Vaishnavi struggled with communicating her ideas and thoughts clearly, which affected her academic performance. Dr Abhinav taught her how to articulate her ideas effectively and present them in a clear and concise manner. He also showed her how to listen actively and communicate with others in a respectful and empathetic manner. With Dr. Abhinav's guidance, Vaishnavi was able to improve her communication skills and perform better in group discussions and presentations.

- **Conflict Management:** Vaishnavi often found it challenging to manage conflicts, especially with her peers. Dr Abhinav taught her how to identify the root cause of conflicts, communicate effectively, and resolve issues in a constructive manner. He also encouraged her to approach conflicts with an open mind and seek to understand the other person's perspective. With Dr Abhinav's guidance, Vaishnavi learned how to manage conflicts effectively and build stronger relationships with her peers.

- **Increasing Influence:** Vaishnavi struggled with gaining influence and respect among her peers and faculty members. Dr Abhinav taught her how to build her reputation by consistently delivering high-quality work, building strong relationships, and taking on leadership roles. He also showed her how to network effectively and build a professional brand. With Dr Abhinav's guidance, Vaishnavi was able to increase her influence and gain respect among her peers and faculty members.

- **Fostering Personal Growth:** Dr Abhinav encouraged Vaishnavi to focus on personal growth and development. He advised her to take on new challenges, seek out new experiences, and pursue her passions. He also helped her identify areas of weakness and develop strategies to overcome them. With Dr Abhinav's guidance, Vaishnavi was able to grow and develop as an individual, both academically and personally.

- **Interacting with people:** Vaishnavi often struggled with interacting with people who were more experienced than her, feeling intimidated and unsure of how to communicate effectively. Dr Abhinav taught her how to approach these interactions with confidence and respect. He advised her to listen actively and ask questions to learn from their experiences. He also encouraged her to seek out mentorship and guidance from experienced professionals. With Dr Abhinav's guidance, Vaishnavi learned how to interact with people who were more experienced than her and gain valuable insights and knowledge from them.

Mentorship in Sports: How Coaches Help Athletes Reach Their Full Potential

The role of mentorship in athletic development cannot be overemphasized. A mentor provides guidance, support, and knowledge to athletes, helping them to reach their full potential. We will explore what makes a good coach and mentor, the benefits of mentorship for athletes, building strong relationships, the importance of setting goals and measuring progress, developing key skills, overcoming challenges and adversity with mentorship, a case study of a mentor-mentee relationship, and lessons learned.

What Makes a Good Coach and Mentor?

A good coach and mentor possess certain qualities that set them apart. First and foremost, a good coach and mentor should be knowledgeable about the sport they are coaching. They should have experience and expertise in the technical, physical, and mental aspects of the sport. In addition to knowledge and experience, a good coach and mentor should be patient, empathetic, and have excellent communication skills. They should be able to explain complex concepts in a way that athletes can understand and be able to adapt their coaching style to the needs of each individual athlete. Finally, a good coach and mentor should lead by example. They should embody the qualities they want to instill in their athletes, such as discipline, determination, and perseverance.

Benefits of Mentorship for Athletes

Mentorship provides a wide range of benefits for athletes. First and foremost, a mentor can provide guidance and support throughout an athlete's career. They can offer advice on training, nutrition, and competition preparation. In addition to guidance and support, a mentor can help athletes to develop key skills, both on and off the field. This includes technical skills, such as technique and strategy, as well as physical and mental skills, such as strength endurance, and focus. Furthermore, mentorship can help athletes to overcome challenges and adversity. A mentor can provide encouragement and motivation during difficult times, helping athletes to stay focused and committed to their goals.

Building Strong Relationships: Communication, Trust, and Respect

Effective mentorship is built on strong relationships between mentors and athletes. This relationship is based on communication, trust and respect. Effective communication is essential for building a strong mentor-mentee relationship. Mentors should be able to listen to their athletes, understand their needs, and provide feedback that is both constructive and supportive. Trust is also critical in any mentor-mentee relationship. Athletes must trust their mentors to provide guidance and support that is in their best interest. Mentors must also trust their athletes to be committed and dedicated to their goals. Finally, respect is essential for building a strong mentor-mentee relationship. Mentors must respect their athlete's individuality and be sensitive to their unique needs and circumstances.

The Importance of Setting Goals and Measuring Progress

Goal setting is a critical component of athletic development. Athletes and mentors should work together to set **s**pecific, **m**easurable, **a**chievable, **r**elevant, and time-bound goals, (SMART goals). These goals should be tailored to the athlete's unique needs and circumstances. Measuring progress is also essential for tracking an athlete's development. Mentors should regularly assess an athlete's performance and adjust their training and development plan accordingly.

Developing Key Skills: Technical, Physical and Mental

Athletic development involves developing key skills, including technical, physical, and mental skills.

Technical skills refer to an athlete's mastery of the specific techniques and strategies required for their sport. Mentors should provide guidance and feedback on technical skills, helping athletes to refine and perfect their techniques. Physical skills refer to an athlete's physical abilities, such as strength, speed, agility, and endurance. Mentors should develop training plans that help athletes to develop and improve their physical skills. Mental skills refer to an athlete's ability to focus, maintain confidence, and manage emotions during competition. Mentors should provide strategies and techniques to help athletes develop mental toughness and resilience, such as visualization, mindfulness, and positive self-talk.

Overcoming Challenges and Adversity with Mentorship

Athletes will inevitably face challenges and adversity throughout their careers. A mentor can provide valuable support and guidance during these times, helping athletes to stay focused and motivated. Mentors can help athletes to develop coping mechanisms and strategies for dealing with setbacks and failures. They can also provide perspective reminding athletes of their long-term goals and helping them to stay positive and focused on the big picture.

Case Study: Mentor-Mentee Relationship

One example Sandeep Singh is a well-known Indian field hockey player who has represented the country at the international level. He is known for his impressive skills his powerful drag-flicking ability became his trademark and he is considered one of the best drag-flickers in the world and leadership on the field. Sandeep's elder brother Bikramjeet Singh played a crucial role in mentoring him and helping him reach his full potential as a field hockey player. Bikramjeet recognized Sandeep's talent at a young age, and powerful drag-flicking ability and started training him in field hockey. He not only helped Sandeep develop his technical skills but also provided him with emotional and mental support. Bikramjeet motivated Sandeep to work hard, set goals, and stay focused on achieving them Hel also taught him the importance of resilience, perseverance and mental toughness.

In 2004, Sandeep made his debut for the Indian national team at **"Sultan Azlan Shah Cup,"** and soon became one of the team's key players. However, in 2006, Sandeep faced a major setback in his career when he was accidentally shot in the spine while travelling on a train to join the Indian team for the World Cup in Germany. The injury left him paralyzed and unable to play field hockey. Sandeep's mentor, Bikramjeet, stepped up to support him during his recovery, encouraging him to stay positive and focused on his rehabilitation. With Bikramjeet's help, Sandeep underwent extensive physiotherapy and rehabilitation, working tirelessly to regain strength and mobility in his legs. He also continued to practice field hockey, using a wheelchair to move around the field. Sandeep's hard work and dedication paid off, and he eventually made a successful comeback to the Indian national team in 2008 just two years after his injury. Throughout Sandeep's career, Bikramjeet continued to provide him with guidance and support, helping him to develop his skills and reach his full potential as a field hockey player. Bikramjeet's mentorship played and crucial role in Sandeep's success, both on and off the field. Their strong bond and commitment to each other demonstrate the value of mentorship in athletic development.

Lessons Learned: How Mentorship Can Help Athletes Reach Their Full Potential

Mentorship is a critical component of athletic development. It provides guidance, support and knowledge to athletes, helping them to reach their full potential. Effective mentorship is based on strong relationships between mentors and athletes, built on communication, trust, and respect. It involves goal setting and measuring progress, as well as the development of key skills, including technical, physical and mental skills. Mentorship can help athletes to overcome challenges and adversity, providing support and guidance during difficult times. By working with a mentor, athletes can develop the skills and mindset they need to achieve their goals and reach their full potential.

Sandeep Singh and his mentor Bikramjeet Singh highlight several important lessons on how mentorship can help athletes reach their full potential:

1. **Importance of having a mentor:** Bikramjeet recognized Sandeep's talent and potential at a young age and decided to mentor him. Having a mentor who

is experienced and knowledgeable can be immensely beneficial for athletes, as they can provide guidance, support and motivation.

2. **Emotional and mental support:** Bikramjeet not only helped Sandeep develop his technical skills but also provided him with emotional and mental support. He motivated Sandeep to work hard, set goals, and stay focused on achieving them. This helped Sandeep develop resilience, perseverance, and mental toughness.

3. **Importance of resilience:** Sandeep faced a major setback in his career when he was accidentally shot in the pine, leaving him paralyzed. However, with Bikramjeet's support, Sandeep remained resilient and persevered through his injury, working tirelessly to regain strength and mobility in his legs.

4. **Overcoming adversity:** Sandeep's injury was a major setback, but with Bikramjeet's help, he was able to overcome it and make a successful comeback to the Indian national team. The experience taught him the importance of overcoming adversity and the power of hard work and dedication.

5. **Building strong relationships:** Bikramjeet and Sandeep's strong bond and commitment to each other demonstrate the importance of building strong relationships in mentorship. Communication, trust, and respect are crucial elements of any successful mentor-mentee relationship.

The Value of Mentorship for Athletes and Coaches Alike

Mentorship is not only beneficial for athletes but also for coaches. Coaches can benefit from mentorship by working with experienced mentors who can provide guidance and support on coaching techniques, communication skills and leadership strategies. By working with a mentor, coaches can refine their coaching skills and develop a deeper understanding of the technical, physical and mental aspects of their sport. This can ultimately lead to improved performance for both coaches and athletes.

Mentorship in Non-Profits: How Mentors Are Making a Difference in Their Communities

Mentorship in non-profits is a powerful tool that can have a significant impact on the development and success of youth and adults. Mentoring programs provide a supportive environment for individuals to learn, grow and achieve their goals with the guidance of a trusted mentor. We will explore the benefits of mentorship programs, the role of non-profits in supporting mentors, evaluating the impact of mentorship programs, and creative approaches to building connections and fostering growth in non-profit settings.

1. **The Importance of Mentorship in Non-Profits: Exploring the Benefits of Mentoring Programs for Youth and Adults**

   Mentorship programs provide numerous benefits to

both youth and adults. For youth, mentorship can help with academic achievement, career preparation, and social-emotional development. It can also provide a positive role model and support system for those who may not have access to these resources at home or in their community. For adults, mentorship can offer career advancement opportunities, professional development, and personal growth. It can also provide a sense of purpose and fulfilment by giving back to their community.

One inspiring example of the impact of mentorship is the story of Big Brothers Big Sisters of America. They have been providing one-to-one mentoring for children facing adversity since 1904. A study conducted by Public/Private Ventures found that youth in the Big Brothers Big Sisters program were 46% less likely to begin using illegal drugs, 27% less likely to begin using alcohol, and 52% less likely to skip school than their peers who were not involved in the program. These statistics demonstrate the power of mentorship in helping youth achieve their goals and avoid risky behaviours.

2. **From Mentee to Mentor: How One Man's Positive Experience with a Non-Profit Led Him to Give Back and Become a Mentor Himself**

Many individuals who have benefited from mentorship programs go on to become mentors themselves, paying it forward and giving back to their community. One example of this is Alphonso Mayo, who was a mentee in the Boys and Girls Club of

America when he was a child. He credits the program and his mentor with helping him overcome obstacles and achieve his goals. As an adult, he became a mentor himself, helping young people in his community through the same program that helped him as a child. He has since been recognized for his contributions to the organization and continues to inspire others to become mentors.

3. **The Role of Non-Profits in Supporting Mentors: Strategies for Recruiting Training, and Retaining Effective Mentors**

Non-profits play a crucial role in supporting mentors by providing training resources and support to help them be effective in their roles. Effective recruitment strategies can help identify potential mentors who are committed to the organization's mission and have the skills needed to be successful. Training programs can help mentors develop the skills and knowledge needed to be effective, such as communication, relationship-building, and problem-solving. Non-profits can also provide ongoing support to mentors through regular check-ins, feedback, and recognition for their contributions.

4. **Measuring Success: Evaluating the Impact of Mentorship Programs on Youth Development, Academic Achievements and Social-Emotional Well-Being**

Evaluating the impact of mentorship programs is

crucial to ensuring they are effective and meet the needs of their participants. Measuring success can include tracking academic achievement, social-emotional well-being, and other outcomes. It can also involve gathering feedback from mentors, mentees, and other stakeholders to assess program effectiveness and identify areas for improvement. Non-profits can use this data to inform program development, improve outcomes and secure funding for future initiatives.

5. Beyond One-on-One Mentoring: Creative Approaches of Building Connections and Fostering Growth in Non-Profit Settings

One-on-one mentoring is a traditional form of mentorship, but there are also creative approaches that non-profits can use to build connections and foster growth in their communities. For example, group mentoring programs can offer a supportive environment for individuals to learn and grow together. Peer mentoring programs can also be effective, where individuals with shared experiences and goals can support each other. Non-profits can also incorporate technology, such as online mentorship platforms, to connect mentors and mentees who may not have the opportunity to meet in person. These creative approaches can offer unique benefits and help non-profits reach a wider audience.

KEY TAKEAWAYS

CHAPTER 4: MENTORS IN THE REAL WORLD: CASE STUDIES AND EXAMPLES OF SUCCESSFUL MENTOR-MENTEE RELATIONSHIPS

Celebrity Mentors: Famous Examples of Mentorship

- Mentor-mentee and with celebrity mentors include the importance of having mentors throughout one's career, creating a supportive and empowering environment for mentees, and leading by example to inspire transformational change. Mentors who can provide guidance, feedback, and support while also allowing their mentees to take ownership of their development can help them reach their full potential. Some famous examples of mentor-mentee relationships include A. P. J Abdul Kalam, Barack Obama, Satya Nadella, Sundar Pichai, Rishi Sunak, Parag Agrawal, Ratan Tata an Narendra Modi. These mentors have instilled in their mentees a strong work ethic, a belief in the power of education, and the importance of giving and helping others.

Mentorship in the Workplace: Success Stories from Corporate America

- Mentorship programs can help develop and retain talent within a company. By pairing high-potential employees with senior leaders, employees can learn from experienced mentors and develop leadership skills.

- Voluntary mentorship programs, such as Intel's "Pay it Forward" program can improve job satisfaction by allowing employees to choose their level of involvement and learn from others.

- Reverse mentorship programs, such as Cisco's program can be effective in keeping senior executives up-to-date with technology trends and market opportunities.

- Global mentorship programs, such as General Electric's program, can help employees develop cross-cultural communication skills and gain a better understanding of global business practices.

- Mentorship programs, such as Microsoft's program, can improve employee retention and engagement by providing opportunities for employees to develop their skills and achieve their goals.

Mentorship in Education: How Teachers and Professors Can Impact Students Lives

- Mentorship from teachers and professors can impact students' lives, using the example of Dr Abhinav Dhotre and his mentee Vaishnavi. This includes the importance of critical thinking, strategic decision-making, problem-solving, setting high expectations, motivating team members, effective teamwork, communication skill, conflict management, increasing influence, fostering personal growth, and interacting with people. Through Dr. Abhinav's guidance, Vaishnavi was able to improve academically, develop life skills and grow personally. This example highlights the significant impact that effective mentorship can have on students' success in and outside of the classroom.

Mentorship in Sports: How Coaches Help Athletes Reach Their Full Potential

- Mentors provide guidance, support, and knowledge to athletes, helping them to reach their full potential.

- Mentors can help athletes develop key skills, both on and off the field, including technical, physical and mental skills

- Mentorship can help athletes to overcome challenges and adversity, providing encouragement and motivation during difficult times.

- Effective mentorship is built on strong relationships between mentors and athletes, based on communication, trust and respect.

- Goal setting and measuring progress and critical components of athletic development.

- A good coach and mentor should be knowledgeable, experienced, patient, empathetic and have excellent communication skills.

- A good coach and mentor should lead by example, embodying the qualities they want to instill in their athletes, such as discipline, determination, and perseverance.

- A good coach and mentor should be able to adapt their coaching style to the needs of each individual athlete.

Mentorship in Non-Profits: How Mentors are Making a Difference in Their Communities

- Mentorship programs can have a significant impact on the development and success of both youth and adults in non-profit settings, providing benefits such as academic achievement, and social-emotional development.

- Non-profits play a crucial role in supporting mentors by providing training resources and support to help them be effective in their roles, and effective recruitment strategies can help identify potential mentors who are committed to the organization's mission.

- Evaluating the impact of mentorship programs is crucial to ensuring they are effective and meet the needs of their participants. Measuring success can include tracking academic achievement, social-emotional well-being, and other outcomes.

- Beyond one-on-one mentoring, non-profits can use creative approaches to build connections and foster growth in their communities, such as group mentoring programs, peer mentoring programs and online mentorship platforms.

- Individuals who have benefited from mentorship programs can become mentors themselves, paying it forward and giving back to their community.

CHAPTER 5: BECOMING A MENTOR: HOW TO PAY IT FORWARD AND HELP OTHERS ACHIEVE THEIR GOALS

> *"As a mentor, you have to be willing to put yourself in your mentee's shoes to understand the struggles that they deal with"* – Toby Keith

Mentoring is a rewarding and fulfilling way to share knowledge, skills, and experience with someone who can benefit from your guidance and support. Whether you are a seasoned professional, a successful entrepreneur, or a passionate hobbyist, you have something valuable to offer to someone who is eager to learn from you. But how do you become a mentor? And what are the benefits of mentoring for both you and your mentee? In this chapter, we will explore some of the key aspects of mentoring, and how you can get started on your mentoring journey.

Why should you become a mentor?

Becoming a mentor can have many benefits for both you and your mentee. Some of the reasons why you should consider becoming a mentor are:

- You can make a positive difference in someone's life by helping them achieve their goals, overcome their challenges, and grow their confidence.

- You can enhance your own skills and knowledge by teaching, explaining, and demonstrating what you know to someone else.

- You can expand your network and build meaningful relationships with people who share your interests, values and passions.

- You can gain satisfaction and fulfilment by giving back to your community and paying it forward to the next generation.

- You can learn new things from your mentee's perspective, experience and feedback.

Becoming a mentor is a rewarding way to pay it forward and help others achieve their goals. It involves sharing your knowledge, skills, and experience to guide and support someone else in their personal or professional development. Here are some brief potential steps to becoming a mentor.

1. **Reflect on your expertise:** Consider your skills, knowledge and experience in a particular field or area where you can provide guidance and support. Reflect on your strengths and areas of expertise that can be valuable to someone seeking mentorship.

2. **Identify potential mentees:** Look for individuals who could benefit from your mentorship. This could be someone in your workplace, community, or even

online through mentorship programs or platforms. Consider their goals, aspirations, and areas where they may need guidance.

3. **Establish clear expectations:** Clarify the scope and goals of the mentoring relationship. Set expectations with your mentee about the frequency and duration of meetings, communication channels and the specific areas in which you will provide support. It's important to have clear boundaries and mutual understanding from the outset.

4. **Build a trusting relationship:** Establish a trusting and respectful relationship with your mentee. Foster open communication, active listening, and empathy. Create a safe space where your mentee feels comfortable sharing their challenges, aspirations and progress.

5. **Provide guidance and support:** Offer guidance and support based on your expertise and experience. Share practical advice, insights, and resources to help your mentee develop new skills, overcome challenges, and achieve their goals. Be patient, encouraging and willing to share both successes and failures from your own journey.

6. **Encourage reflection and growth:** Encourage your mentee to reflect on their progress, set goals and develop strategies for their personal and professional growth. Help them identify their strengths, weaknesses, opportunities, and challenges and

support them in developing strategies to address them.

7. **Foster independence:** Encourage your mentee to become independent and self-reliant by gradually empowering them to make their own decisions and take ownership of their goals. Your role as a mentor is to guide, not to dictate or impose your opinions.

8. **Continuously evaluate and adjust:** Continuously evaluate the progress of your mentee and the effectiveness of the mentoring relationship. Adjust your approach, if needed based on their evolving needs and goals.

9. **Celebrate achievements:** Celebrate the achievements and milestones of your mentee. Recognize their progress and offer encouragement and positive feedback to motivate them to continue striving for their goals.

10. **Pay it forward:** Encourage your mentee to pay it forward by becoming a mentor themselves and helping others in their journey. Mentoring is a cycle of continuous learning and growth, and encouraging your mentee to become a mentor themselves can create a ripple effect of positive impact.

Becoming a mentor can be a transformative experience both for the mentee and the mentor. It's a way to give back, share your knowledge and experience and make a meaningful difference in someone's life. Remember that mentoring is a two-way relationship that requires commitment, patience and empathy. By helping others achieve their goals, you can create a positive impact that extends far beyond the immediate sphere of influence.

The Benefits of Being a Mentor: How Mentorship Can Help You Grow

Mentorship is a valuable practice that can benefit both the mentor and the mentee. By sharing your knowledge, skills and experience with someone who is eager to learn, you can help them achieve their goals and grow as a person. But mentorship is not a one-way street. As a mentor, you can also gain a lot from the relationship. Here are some of the benefits of being a mentor:

- You can expand your network and perspective. By mentoring someone from a different background, industry, or field, you can broaden your horizons and learn new things. You can also build connections with other mentors and mentees, and access new opportunities and resources.

- You can enhance your self-esteem and satisfaction. By seeing the positive impact, you have on your mentee's life, you can boost your confidence and self-worth. You can also feel a sense of fulfilment and purpose by giving back to your community and profession.

- You can grow personally and professionally. By reflecting on your own journey, strengths, and weaknesses, you can gain new insights and perspectives on yourself. You can also challenge yourself to learn new skills, update your knowledge, and pursue new goals.

As the famous author and speaker Zig Ziglar once said, *"A lot of people have gone further than they thought they could because someone else thought they could."* Being a mentor can help you unlock your own potential as well as that of your mentee.

- You can develop leadership and communication skills. Mentoring requires effective communication, active listening and the ability to provide constructive feedback. As a mentor, you can improve your communication and leadership skills, which can be valuable in various aspects of your personal and professional life.

- You can gain a fresh perspective and new ideas. Mentees often bring fresh perspectives and innovative ideas to the table. By engaging in discussions and brainstorming sessions with your mentee, you can gain new insights and challenge your own assumptions. This can spark creativity and help you approach challenges with a fresh mindset.

- You can leave a lasting legacy. As a mentor, you have the opportunity to make a positive impact on someone's life and career. By sharing your knowledge, experience and wisdom you can help shape the future of your mentee and leave a lasting legacy in their development. This can bring a sense of fulfilment and pride in your contributions to the next generation.

- You can gain personal satisfaction and happiness. Helping someone else succeed and achieve their goals can be deeply rewarding. It can bring a sense of joy, satisfaction, and happiness knowing that you have made a positive impact on someone's life. This can contribute to your overall well-being and happiness, enhancing your personal growth and fulfilment.

Mentorship is a mutually beneficial relationship that can provide numerous benefits to both the mentor and the mentee. As a mentor, you can expand your network, enhance your self-esteem, grow personally and professionally, develop leadership and communication skills, gain fresh perspectives, leave a lasting legacy, stay updated in your field and experience personal satisfaction and happiness. By investing your time and expertise in mentoring others, you can unlock your own potential and contribute to the growth and development of others.

Qualities of a Good Mentor: What it Takes to Be an Effective Mentor

Being an effective mentor requires a diverse range of qualities and skills. Here are some key qualities that are typically associated with a good mentor.

1. **Experience and Expertise:** A good mentor should have significant experience and expertise in the field or area they are mentoring in. They should have a deep understanding of the subject matter and be able to provide guidance and insights based on their own experiences.

2. **Excellent Communication Skills:** Communication is a crucial aspect of mentoring. A good mentor should be able to communicate effectively, both verbally and in writing, and should be able to listen actively and empathetically. They should be able to explain complex concepts in a simple and understandable manner and provide feedback in a constructive and encouraging manner.

3. **Patience and Empathy:** Mentoring can be a challenging and time-consuming process. A good mentor should be patient and understanding and should be able to empathize with the mentee's challenges and concerns. They should be able to create a supportive and inclusive environment that encourages open communication and fosters trust.

4. **Passion for Teaching and Sharing Knowledge:** A good mentor should have a genuine passion for teaching and sharing their knowledge with others. They should be enthusiastic about helping others succeed and be willing to invest time and effort in the mentoring relationship.

5. **Ability to Inspire and Motivate:** A good mentor should be able to inspire and motivate mentees to

achieve their full potential. They should be able to set high standards and expectations, while also providing the necessary encouragement and support to help their mentees overcome challenges and obstacles.

6. **Flexibility and Adaptability:** Every mentee is unique, and a good mentor should be able to adapt their mentoring style to the needs and learning style of each mentee. They should be flexible and open-minded and be able to adjust their approach as needed to best support the mentee's growth and development.

7. **Trustworthiness and Confidentiality:** A good mentor should be trustworthy and maintain confidentiality. Mentees should feel comfortable sharing their concerns, challenges, and aspirations with their mentor, and trust that their mentor will keep their conversations confidential unless there is a legal or ethical obligation to disclose information.

8. **Positive Role Modelling:** A good mentor should lead by example and demonstrate professionalism, integrity, and ethical behavior. They should serve as a positive role model for their mentees, showcasing the qualities and behaviors that lead to success in their field or area of expertise.

9. **Continuous Learning and Growth Mindset:** A good mentor should be committed to their own continuous learning and growth. They should stay updated with the latest developments in their field and be willing to learn from their mentees as well. They should also be open to receiving feedback and be

willing to reflect on their own mentoring practices to improve and enhance their effectiveness as a mentor.

10. **Genuine Care and Support:** Lastly, a good mentor should genuinely care about the well-being and success of their mentees. They should be invested in their mentees' personal and professional growth and be willing to provide guidance, support, and encouragement throughout the mentoring relationship.

Finding Mentees: How to Identify and Connect with Potential Mentees

Finding potential mentees is an important step in the mentoring process. Identifying mentees who could benefit from your guidance and expertise can lead to meaningful mentor-mentee relationships. Here are some tips on how to identify and connect with potential mentees:

- **Be Visible and Accessible:** Make yourself visible and accessible to potential mentees. Share your expertise and knowledge in relevant forums, events or online platforms related to your field or area of expertise. This can include professional associations, industry events, networking events, or online communities. Be approachable and open to conversations and be willing to share your insights and experiences.

- **Seek Out Mutual Interests:** Look for mentees who share mutual interests or aspirations in your field or area of expertise. This can help create a strong foundation for a meaningful mentoring relationship. You can identify potential mentees through professional networks, referrals, or by reaching out to individuals who express an interest in your work or industry.

- **Be Proactive in Offering Help:** Take the initiative to offer your help and guidance to potential mentees. This can be done through informal conversations, informational interviews or by reaching out to individuals who you believe could benefit from your mentorship. Show genuine interest in their career goals and aspirations and offer to share your knowledge and insights.

- **Look for a Willingness to Learn:** Seek mentees who demonstrate a willingness to learn and grow. Look for individuals who are curious, motivated, and eager to develop their skills and knowledge. A mentee who is proactive and receptive to feedback is more likely to benefit from your mentorship.

- **Foster Authentic Connections:** Building a meaningful mentor-mentee relationship requires authentic connections. Look for potential mentees with whom you share common values, and whose personality and work ethic align with you own. This can help establish a strong foundation for a successful mentoring relationship built on trust and respect.

- **Utilize Mentoring Programs or Platforms:** Many organizations or institutions have formal mentoring programs or platforms that can help you connect with potential mentees. These programs often match mentors and mentees based on their interests, skills, and goals, making it easier to identify and connect with potential mentees.

- **Be Patient and Persistent:** Finding the right mentees may take time and effort. Be patient and persistent in your search for potential mentees. Keep networking, attending relevant events, and reaching out to individuals who could benefit from your mentorship. Building meaningful mentor-mentee relationships may require time and perseverance.

Providing Support and Feedback: How to Guide Your Mentee to Success

As a mentor, your role is crucial in guiding your mentee towards success. While offering support and feedback may seem like intuitive tasks, there are specific strategies that can enhance your effectiveness in this role. In this chapter, we will explore actionable steps that you can take to provide meaningful support and feedback to your mentee, ultimately empowering them to achieve their goals.

1. **Establish Clear Expectations:** Begin by setting clear expectations with your mentee. Clearly define the goals, objectives, and timelines for the mentoring

relationship. Ensure that your mentees understand what they can expect from you as their mentor, as well as what you expect from them in terms of commitment and effort. This will provide a solid foundation for effective support and feedback throughout the mentoring relationship.

2. **Build Trust:** Trust is the cornerstone of any successful mentoring relationship. It is important to establish a trusting relationship with your mentee early on. Show genuine interest in their goals and aspirations, listen actively to their concerns and maintain confidentiality. Be reliable, consistent and honest in your interactions with your mentee. Building trust will create a safe space for your mentees to share their thoughts, seek feedback, and take risks.

3. **Provide Constructive Feedback:** Feedback is a critical component of mentoring. When providing feedback, be specific, objective, and constructive. Avoid criticizing or judging your mentee, as this can discourage them and hinder their progress. Instead, focus on providing feedback that highlights their strengths, identifies areas for improvement, and offers suggestions for growth. Frame feedback as an opportunity for learning and development rather than criticism.

4. **Use Effective Communication Skills:** Effective communication is key to guiding your mentee to success. Practice active listening, which involves being fully present, maintaining eye contact, and

avoiding interruptions. Clarify any misunderstandings, ask open-ended questions to encourage reflection and provide feedback in a respectful and non-judgmental manner. Use positive language and tone and be mindful of your body language and non-verbal cues. Your communication style can significantly impact the effectiveness of your support and feedback.

5. **Foster Self-Reflection:** Encourage your mentee to engage in self-reflection. Help them develop the habit of regularly reflecting on their progress, strengths, areas for improvement and future goals. Ask thought-provoking questions that prompt critical thinking and self-assessment. Encourage your mentee to set their own goals and develop action plans to achieve them. By fostering self-reflection, you are empowering your mentee to take ownership of their growth and development.

6. **Offer Resources and Opportunities:** Provide your mentee with resources, opportunities, and networks that can support their growth and development. Share relevant articles, books, websites, and tools that can enhance their knowledge and skills. Connect them with professionals in their field of interest, and encourage them to participate in workshops, conferences, or other relevant events. Offer guidance on how to navigate challenges, make decisions, and seize opportunities. Providing access to resources and opportunities can significantly contribute to your mentee's success.

7. **Celebrate Achievements:** Acknowledge and celebrate your mentee's achievements, no matter how small they may seem. Recognize their efforts, progress and accomplishments. Celebrating achievements boosts your mentee's confidence, motivation and self-esteem. It also reinforces positive behaviour and encourages continued growth and development.

Ending the Mentorship Relationship: How to Wrap Up a Successful Mentorship

As a mentor, the end of a mentoring relationship can be bittersweet. While it's a time to reflect on the progress and growth of your mentee, it's also important to ensure a smooth and successful conclusion to the mentorship. In this chapter, we will explore actionable steps that can help you effectively end a mentorship relationship and leave a positive impact on your mentee.

- **Review and Reflect:** Take the time to review and reflect on the goals, objectives and milestones that were set at the beginning of the mentorship. Evaluate the progress that your mentee has made towards achieving their goals and celebrate their achievements. Reflect on the lessons learned, challenges overcome, and the overall impact of the mentorship on your mentee's growth and development.

- **Set a Timeline:** Once you have reviewed and reflected on the mentorship, set a timeline for wrapping up the relationship. This timeline can be based on the original timeframe that was established at the beginning of the mentorship or can be adjusted to align with your mentee's progress and needs. Communicate the timeline clearly to your mentee and ensure that you both are on the same page.

- **Recap and Summarize:** Before concluding the mentorship, recap and summarize the key takeaways, insights and learnings from the entire mentoring journey. Provide your mentee with a comprehensive summary of the progress they have made, the skills they have developed, and the achievements they have accomplished. This recap can serve as a reference for your mentee as they continue their professional journey.

- **Evaluate Mentee's Readiness for Independence:** Assess your mentee's readiness for independence and determine if they are capable of continuing their professional growth without your guidance. Evaluate their level of confidence, skills, and knowledge in their field of interest. If you feel that your mentees are ready to stand on their own, have an open conversation with them about their readiness and provide them with feedback and encouragement.

- **Develop a Transition Plan:** Work with your mentee to develop a transition plan that outlines their next steps after the mentorship ends. This plan can include goals they plan to pursue, strategies for maintaining momentum, and resources they can rely on moving forward. Offer suggestions and guidance on how to continue their growth journey independently and encourage them to take ownership of their career development.

- **Celebrate and Express Gratitude:** End the mentorship relationship on a positive note by celebrating the accomplishments of your mentee and expressing gratitude for the opportunity to mentor them. Acknowledge their efforts, progress and achievements and provide specific feedback on the positive impact they have had on you as a mentor. Express your appreciation for the trust, commitment and collaboration that you both have shared throughout the mentorship.

KEY TAKEAWAYS

CHAPTER 5: BECOMING A MENTOR: HOW TO PAY IT FORWARD AND HELP OTHERS ACHIEVE THEIR GOALS

The Benefits of Being a Mentor: How Mentorship Can Help You Grow

1. Mentoring can be a rewarding and fulfilling way to share your knowledge skills and experience with someone who can benefit from your guidance and support.

2. As a mentor, you can make a positive difference in someone's life, enhance your own skills and knowledge, expand your network, gain satisfaction and fulfilment, and learn new things from your mentee's perspective.

3. To become a mentor, you should reflect on your expertise, identify potential mentees, establish clear expectations, build a trusting relationship, provide guidance and support, encourage reflection and growth, foster independence, continuously evaluate and adjust, celebrate achievements and pay it forward.

4. Mentoring is a two-way relationship that requires commitment, patience and empathy. It's a cycle of continuous learning and growth that can create a positive impact that extends far beyond the immediate sphere of influence.

5. By becoming a mentor, you have the opportunity to give back to your community and pay it forward to the next generation, while also developing your own skills, knowledge and network.

Qualities of a Good Mentor: What It Takes to Be an Effective Mentor

1. Experience and expertise in the field they are mentoring in.
2. Excellent communication skills, including active listening and constructive feedback.
3. Patience and empathy towards the mentee's challenges and concerns.
4. Passion for teaching and sharing knowledge.
5. Ability to inspire and motivate mentees.
6. Flexibility and adaptability in their mentoring style.
7. Trustworthiness and maintaining confidentiality.
8. Positive role modelling through professionalism, integrity, and ethical behavior.
9. Commitment to continuous learning and growth mindset.
10. Genuine care and support for the well-being and success of mentees.

Finding Mentees: How to Identify and Connect with Potential Mentees

1. Be visible and accessible: Make yourself known and available in relevant forums, events or online platforms related to your field or area of expertise to attract potential mentees.

2. Seek out mutual interest: Look for mentees who share common interests or aspirations in your field, as this can create a strong foundation for a meaningful mentoring relationship.

3. Be proactive in offering help: Take the initiative to offer your guidance and expertise to potential mentees through informal conversations, informational interviews or by reaching out to individuals who could benefit from your mentorship.

4. Look for a willingness to learn: Seek mentees who are curious, motivated and eager to learn and grow, as they are more likely to benefit from your mentorship.

5. Foster authentic connections: Building a meaningful mentor-mentee relationship requires authentic connections based on shared values, personality, and work ethic.

6. Utilize mentoring programs or platforms: Take advantage of formal mentoring programs or platforms offered by organizations or institutions to connect with potential mentees who align with your interests, skills and goals.

7. Be patient and persistent: Finding the right mentees may take time and effort, so be patient and persistent in your search, and continue networking, attending events and reaching out to potential mentees who could benefit from your mentorship.

Providing Support and Feedback: How to Guide Your Mentee to Success

1. Setting clear expectations at the beginning of the mentoring relationship can provide a foundation for effective support and feedback.

2. Building trust is essential to create a safe space for your mentee to seek feedback and take risks.

3. When providing feedback, be specific, objective and constructive and frame it as an opportunity for learning and development rather than criticism.

4. Effective Communication, including active listening and using positive language and tone, can significantly impact the effectiveness of your support and feedback.

5. Fostering self-reflections and encouraging your mentee to take ownership of their growth and development can empower them to achieve their goals.

6. Offering resources, opportunities and networks can support your mentee's growth and development.

7. Celebrating achievements can boost your mentee's confidence, motivation and self-esteem, and encourage continued growth and development.

Ending the Mentorship Relationship: How to Wrap Up a Successful Mentorship

1. It's important to review and reflect on the progress and growth of your mentee before ending the mentorship.

2. Setting a timeline for wrapping up the relationship and communicating it clearly to your mentee is crucial.

3. Before concluding the mentorship, recap and summarize the key takeaways, insights, and learnings from the entire mentoring journey.

4. It's important to assess your mentees' readiness for independence and determine if they are capable of continuing their professional growth without your guidance.

5. Developing a transition plan with your mentees can help them outline their next steps after the mentorship ends.

6. Ending the mentorship relationship on a positive note by celebrating your mentee's accomplishments and expressing gratitude for the opportunity to mentor them can leave a lasting positive impact.

CHAPTER 6: THE PSYCHOLOGY OF MENTORSHIP: UNDERSTANDING THE DYNAMICS OF MENTOR-MENTEE RELATIONSHIPS

> *"Effective mentorship requires more than just imparting knowledge and experience; it demands empathy, active listening, and a willingness to learn and grow alongside your mentee."* – Adam Grant

Mentorship is an integral part of human development that has played a significant role in the success of individuals throughout history. Mentor-mentee are essential for growth, learning and progress in various aspects of life, from career development to personal well-being. By examining the psychology behind these relationships, we can better understand their dynamics and harness their full potential. Empathy is a critical factor in the effectiveness of mentor-mentee relationships. A successful mentor must have the ability to empathize with their mentee, enabling them to form a deeper connection and better understand the individual's needs, aspirations, and challenges. This level of understanding promotes trust and open communication, creating a strong foundation for working together towards shared goals. Studies consistently show that mentorship characterized by empathy and emotional intelligence results in more impactful and long-lasting positive outcomes for both parties involved.

Another crucial component of effective mentorship is motivation. People often seek mentors to help them navigate unfamiliar territory or overcome obstacles on their path to success. As a result, it is essential for mentors to adopt motivational tactics that inspire their mentees to strive for personal excellence and growth. This can be achieved through goal-setting practices or employing constructive feedback designed to encourage continuous self-improvement. When mentees feel inspired to push their boundaries and tackle challenges head-on with unwavering enthusiasm, they are more likely to achieve their objectives with the guidance and support of their mentor. Additionally, cognitive congruence plays an integral role in successful mentorship dynamics. Cognitive congruence refers to the alignment between the thought processes and learning styles adopted by both mentors and mentees. When mentors adopt teaching methods tailored to the cognitive styles of their mentees. When mentors adopt teaching methods tailored to the cognitive styles of their mentees, they are better equipped to facilitate learning experiences that resonate with everyone's unique approach to processing information. This personalized approach fosters a sense of understanding and appreciation for diverse learning styles, ensuring mentees receive the support and guidance they require to excel in their respective pursuits.

Moreover, a mentor's ability to provide timely feedback is crucial to the mentee's growth and development.

Feedback is an opportunity for the mentees to reflect on their progress, identify areas of improvement and develop a plan to address them. Mentors must deliver feedback constructively and in a manner that is specific, timely and actionable. When feedback is given correctly it motivates mentees to strive for excellence and take ownership of their growth. Furthermore, mentorship is an opportunity for mentors to learn and grow alongside their mentees. As mentors guide their mentees through various challenges and experiences, they gain valuable insights into their mentees' perspectives and develop a deeper understanding of their own strengths and limitations. Mentors can use these insights to refine their approaches, learn new skills, and gain new perspectives, improving their ability to support and mentor others. Mentorship is a powerful tool that has the potential to transform lives and cultivate unwavering success. By examining the underlying psychology of these relationships, we can better understand the dynamics at play between mentors and mentees. Incorporating empathy, motivation, cognition, congruence and timely feedback into mentorship practices enables tailored support to one another on our individual journeys towards personal and professional growth. As we continue to explore the psychology of mentorship, we can unlock its full potential and create more meaningful and impactful mentor-mentee relationships.

The Psychological Benefits of Mentorship for Mentees

Mentorship has long been recognized as an essential component of personal and professional growth. Mentees have the opportunity to learn from experts in their field, soak up invaluable insight, and develop strong networks. Yet, the benefits of mentorship often extend beyond these practical elements. Specifically, this journey offers a plethora of psychological benefits, from increased self-confidence to better decision-making abilities.

- **Difficulty Navigating Professional Challenges:** Every individual faces challenges throughout their career. For mentees, these difficulties can seem insurmountable without adequate support and guidance. Mentors offer personalized advice based on their own experiences, ensuring that mentees never have to face hardship alone.

- **Seek out a Mentor within Your Industry:** Taking the initiative to find a mentor with experience in your field can alleviate feelings of being lost or overwhelmed by professional hurdles. A trusted mentor can offer practical solutions and encouragement to navigate even the most challenging situations.

- **Lack of Confidence and self Confidence and Self-Doubt:** Many people encounter moments where they doubt their abilities, hindering personal and professional growth. Moreover, this lack of confidence can lead to a negative self-image and reduced satisfaction in life.

- **Leverage Your Mentor's Expertise to Foster Self-Confidence:** Establishing a relationship with a mentor enables mentees to gain valuable feedback on their strengths and areas for improvement. This honest dialogue allows for the development of self-confidence as individuals recognize their potential for success with guidance from their mentors.

- **Stephanie's Transformation through Mentorship:** Stephanie had always dreamed of becoming an accomplished architect but faced ongoing challenges and pressure at work, which led her to doubt her abilities. When she began working under Linda, an experience architect who took Stephanie under her wing, thing started to improve. Linda provided Stephanie with constructive feedback and taught her how to approach projects more effectively, which boosted her self-confidence. Through mentorship, Stephanie not only became better architect but also learned to believe in herself and her capabilities.

- **The Lasting Impact of Mentorship:** The psychological benefits of mentorship for mentees are profound and transformative. By addressing pain points such as navigating professional challenges and boosting self-confidence, mentorship equips individuals with the tools to succeed in their chosen careers. Engaging in this relationship empowers mentees to face the future with confidence, knowing that they have the support, guidance and resources needed for achievement.

- **The Psychological Benefits of Mentorship for Mentors**

Krishna is a highly successful business coach who has helped numerous entrepreneurs achieve their goals.

Among his mentees are Arjun, Radha and Lakshmi all of whom are mid-level entrepreneurs with families to support. Arjun is married with two children; Radha is married with a one-year-old baby girl and Lakshmi is a single mother with a five-year-old boy. Krishna has been a mentor for several years and has seen the impact of his guidance on his mentees'. However, he has also experienced personal benefits from the mentor-mentee relationship. When he first started mentoring, he was unsure of his abilities and felt that he lacked the necessary experience to guide others. Krishna struggled with self-doubt and imposter syndrome, wondering if he was truly qualified to be a mentor. This lack of confidence made it challenging for his to give his mentees the support they needed, as he was constantly second-guessing himself. Krishna realized that his doubts were holding him back from fully committing to the mentor-mentee relationship. He recognized that he needed to work on his own mindset and seek support from other mentors to overcome his doubts and become a more effective mentor.

Krishna sought guidance from other mentors in his field, who provided him with valuable feedback and helped him see his strengths. He also began to prioritize his own personal growth and development, attending workshops and seminars to improve his coaching skills. Over time, Krishna's confidence grew, and he was able to provide more meaningful guidance to his mentees. He also gained a deeper understanding of himself and his strengths, which allowed him to better serve his clients.

Personal and Professional Growth Through Mentorship

As a mentor, Krishna has helped his mentees achieve their personal and professional goals. However, he has also experienced significant growth in his own life through the mentor-mentee relationship. Krishna had reached a point in his career where he felt like he had hit a plateau. He was successful but felt unfulfilled and lacked direction. Krishna recognized that he needed to challenge himself and push past his comfort zone if he wanted to continue growing. He also recognized that he needed to seek out new experiences and perspectives to gain a fresh outlook on his work. Krishna began mentoring others, which allowed him to gain a new perspective on his work and discover new opportunities for growth. He also sought out mentorship from other successful coaches, who provided him with guidance and feedback on his coaching style. Through these experiences, Krishna was able to develop new skills, gain confidence, and find renewed purpose in his work.

Developing Leadership and Coaching Skills as a Mentor

Krishna has developed a reputation as an expert in his field and has helped many entrepreneurs achieve success. However, he recognized that he still had room for growth as a mentor and wanted to continue developing his leadership and coaching skills. Krishna recognized that he had developed some bad habits as a mentor, including micromanaging his mentees and not giving them enough autonomy to make their own decisions.

Krishna realized that these habits were holding his mentees back and limiting their growth. He recognized that he needed to develop new coaching skills to help his mentees become more self-sufficient and confident in their own abilities. Krishna attended workshops and seminars to learn new coaching techniques and develop his leadership skills. He also sought feedback from his mentees to understand their needs and challenges better. Through this process, Krishna was able to develop a more effective coaching style that empowered his mentees to take ownership of their own success. He learned to give his mentees more autonomy while still providing guidance and support, which helped them build their own confidence and decision-making abilities.

Building a Sense of Community and Belonging Through Mentorship

Krishna recognized the importance of building a sense of community and belonging among his mentees. Arjun, Radha, and Lakshmi were all mid-level entrepreneurs who faced unique challenges in their personal and professional lives. Krishna noticed that his mentees often felt isolated and alone in their struggles, which made it challenging for them to stay motivated and focused on their goals. Krishna realized that he needed to create a more supportive and collaborative environment among his mentees. He recognized that by building a sense of community and belonging, his mentees would be more likely to succeed and thrive in their businesses. Krishna organized regular meetings and events for his mentees to connect and share their experiences. He also encouraged them to collaborate on projects and support each other in their personal and professional lives. Through these efforts, Krishna was able to create a strong sense of community among his mentees, which help them feel more connected and supported in their entrepreneurial journeys.

Finding Purpose and Fulfilment Through Mentoring Others

Krishna found great purpose and fulfilment in mentoring others. He recognized the impact that he was having on his mentees' lives and felt a deep sense of satisfaction in helping them achieve their goals. Krishna had struggled with finding purpose and meaning in his work before he started mentoring. However, through his experience as a mentor, he found a renewed sense of purpose and passion for his work. Krishna recognized that he needed to find a way to make a meaningful impact in the world through his work. He wanted to feel like he was making a difference and helping others achieve their dreams. Krishna found that mentoring others was a powerful way to make a positive impact in the world. Through his work as a mentor, he was able to help others achieve their goals and make a difference in their own lives. He also found that mentoring gave him a renewed sense of purpose and passion for his work, which helped him feel more fulfilled and satisfied in his career.

Cultivating a Positive Impact and Legacy Through Mentoring

Krishna recognized the importance of leaving a positive impact and legacy through his work as a mentor. He wanted to make a meaningful difference in the world and leave a lasting legacy that would inspire others. Krishna struggled with feeling like he was making a meaningful impact in the world. He wanted to leave a positive legacy but didn't know where to start. Krishna recognized that he needed to find a way to make a difference through his work as a mentor. He wanted to cultivate a positive impact that would inspire others and leave a lasting legacy. Krishna focused on mentoring others with a strong sense of purpose and passion. He recognized that by helping others achieve their goals, he was making a positive impact in the world and leaving a lasting legacy. He also focused on creating a culture of mentorship, where his mentees were inspired to pay it forward and mentor others themselves. Through these efforts, Krishna was able to cultivate a positive impact and legacy that inspired others to follow in his footsteps.

Common Mentor-Mentee Relationship Patterns and How to Navigate Them

1. **Importance of Boundaries:** Although building a strong relationship with a mentee is important, it is equally important to maintain professional boundaries. Mentors should avoid getting too emotionally involved in their mentees' personal lives, as it can lead to confusion and potential conflicts of interest.

2. **Power Dynamics:** Power dynamics can often be present in mentor-mentee relationships, as mentors

may have more experience and authority than their mentees. This can make mentees hesitant to speak up or challenge their mentor's ideas. It is important for mentors to create a safe space for open communication and encourage mentees to express their thoughts and opinions.

3. **Conflict Resolution:** Conflicts can arise in any relationship, and mentor-mentee relationships are no exception. Mentors should be prepared to address conflicts with their mentees and work towards finding a resolution that is beneficial for both parties. Conflict resolution skills are essential in building strong and lasting relationships.

4. **Diversity and Inclusivity:** It is important for mentors to understand and appreciate diversity in all its forms. Mentors should be mindful of cultural, social, and personal differences that may impact their mentees' experiences. A lack of understanding or appreciation for diversity can lead to a breakdown in communication and potentially harm the mentor-mentee relationship.

5. **Accountability:** Mentors should hold themselves accountable for the guidance and advice they provide to their mentees. It is important for mentors to take ownership of their mistakes and work towards finding solutions to any problems that may arise. This level of accountability helps to build trust and respect in the mentor-mentee relationship.

6. **Flexibility:** Every mentee is different, and mentorship should be tailored to their individual needs and goals. Mentors should be flexible and willing to adjust their approach based on their mentees' feedback and progress. This flexibility helps to ensure that the mentor-mentee relationship remains productive and effective over time.

The Role of Trust in Mentorship Relationships

Samantha had been working at her company for five years when she was promoted to the position of mid-level manager. It was a big set-up for her, and she was excited to take on new responsibilities and challenges. However, she quickly realized that she was going to need some help navigating her new role. That's when Samantha met her mentor, Maria. Maria had been with the company for over a decade and had worked her way up to a senior management position. Samantha was thrilled to have the opportunity to learn from someone with so much experience and knowledge. As Samantha and Maria began working together, Samantha realized that one of the most important aspects of their mentorship relationship was trust. She needed to be able to trust Maria to give her honest feedback and guidance, even when it wasn't what she wanted to hear.

Over time, Samantha learned to rely on Maria for support and advice. Maria was always available to talk through challenges and help Samantha develop strategies for overcoming them. Samantha appreciated that Maria was a straight shooter, never sugar-coating her feedback or opinions. She knew that she could always count on Maria to tell her the truth, even when it was difficult to hear. As Samantha continued to grow and develop in her role as a manager, she realized that her relationship with Maria had become one of the most valuable assets in her professional life. She knew that she could always turn to Maria for guidance and support and that Maria had her best interests at heart.

Ultimately, Samantha realized that the key to a successful mentorship relationship was trust. Without trust, she wouldn't have been able to fully engage in the learning and development process, and she wouldn't have been able to make the most of her mentor's expertise and guidance. But with trust, she was able to grow and thrive in her role as a manager and develop skills and insights that would serve her well throughout her career.

Common Barriers to Effective Mentorship Relationships

Arjun was recently promoted to a mid-level manager role where he was responsible for leading a team. Arjun had the technical skills and expertise required for the role but lacked the leadership skills needed to manage and lead his team effectively. To help him overcome those challenges, Krishna, a senior-level manager with excellent leadership skills, became Arjun's mentor.

Mismatched Expectations: Arjun expected that he would receive clear instructions and guidance from

Krishna on how to lead his team effectively. However, Krishna had a different approach to mentoring, which involved guiding Arjun to find his own solutions. This caused confusion and frustration for Arjun. Arjun felt overwhelmed and unsure of how to proceed with leading his team. He became frustrated that he was not receiving clear instructions from his mentor and felt like he was not making any progress. Krishna realized that Arjun had a different expectation of mentoring than he did. To address this issue, Krishna scheduled a meeting with Arjun to discuss their mentoring expectations and align their goals. This helped Arjun understand that Krishna's approach was more about empowering him to find his own solutions rather than providing him with a clear-cut plan.

Lack of Communication: Arjun was not communicating effectively with his team, which was leading to misunderstandings and mistakes. However, he was hesitant to reach out to Krishna for advice on how to communicate effectively. Arjun was struggling to communicate effectively with his team, which was causing confusion and leading to mistakes. He felt like he did not have the necessary skills to address this issue, and he was hesitant to reach out to Krishna for help. Krishna noticed that Arjun was struggling with communication and encouraged him to ask for help. Krishna suggested that Arjun should schedule regular team meetings and encourage open communication. Krishna also provided Arjun with communication training to help him improve his communication skills.

Limited Availability and Commitment: Arjun had a busy schedule and was not always available to meet with Krishna. This was leading to delays in mentorship sessions, and Arjun was not making progress as quickly as he wanted to. Arjun was frustrated that he was not able to meet with Krishna as often as he would have liked. He felt like this was slowing down his progress, and he was worried about falling behind. Krishna recognized that Arjun had a busy schedule and worked with him to find a suitable meeting time. They also set specific goals for each mentoring session to ensure that they were making progress, even if they could meet as frequently.

Power Dynamics and Hierarchical Structures: Arjun felt intimidated by Krishna's seniority and was hesitant to express his opinions or ideas. This was hindering his ability to lead his team effectively. Arjun felt like he could not express his ideas or opinions freely around Krishna. He felt like he was not being heard or taken seriously, which was impacting his confidence and ability to lead his team. Krishna recognized that there was a power dynamic at play and worked to create a safe and open environment for Arjun to express his ideas. Krishna also encouraged Arjun to take ownership of his ideas and to present them with confidence.

Resistant to Change and Feedback: Arjun was resistant to change and feedback, which was hindering his ability to improve his leadership skills. Arjun was resistant to change and feedback, which was causing him to stagnate in his development. He felt like he was not making any progress and was frustrated with his lack of growth. Krishna recognized that Arjun was resistant to change and feedback and worked to create a safe environment for him to receive constructive criticism. Krishna also encouraged Arjun to embrace change and to see feedback as an opportunity for growth.

Cultural Differences and Biases: Arjun and Krishna came from different cultural backgrounds, which impacted their communication and understanding of certain issues. Arjun felt like he was not being understood by Krishna, and he was struggling to understand some of Krishna's perspectives. He was worried that their cultural differences would hinder their mentoring relationship. Krishna recognized the cultural differences and worked to understand Arjun's perspective better. He also provided cultural sensitivity training to help Arjun understand some of Krishna's cultural perspectives.

Personality Conflicts and Communication Styles: Arjun and Krishna had different personality types and communication styles, which was leading to misunderstandings and conflicts. Arjun felt like he was not being understood by Krishna, and their communication styles were clashing. He was worried that their personality conflicts would hinder their mentoring relationships. Krishna recognized the personality conflicts and worked to understand Arjun's communication style better. He also provided communication style training to help Arjun understand some of Krishna's communication preferences.

Lack of Trust and Respect: Arjun was struggling to earn the trust and respect of his team, which was impacting his ability to lead effectively. Arjun felt like he was not being respected by his team, which was causing him to doubt his abilities. He was worried that his lack of trust and respect would hinder his team's performance. Krishna recognized that Arjun was struggling with trust and respect issues and provided him with guidance on how to earn the respect of his team. Krishna also encouraged Arjun to be authentic and vulnerable with his team to help build trust.

Inadequate Training and Resources: Arjun felt like he did not have the necessary training and resources to lead his team effectively. Arjun felt like he was not equipped with the necessary skills and resources to lead his team effectively. He was worried that this would impact his team's performance. Krishna recognized that Arjun needed additional training and resources and provided him with access to leadership training programs and resources. Krishna also encouraged Arjun to seek out additional training and resources on his own.

Time Constraints and Competing Priorities: Arjun had a busy schedule, and his mentorship sessions with Krishna were sometimes postponed or cancelled due to competing priorities. Arjun felt like he did not have enough time to devote to these mentorship sessions, and he was worried that this would hinder his progress. Krishna recognized that Arjun had a busy schedule and worked with him to find a suitable meeting time. They also set specific goals for each mentoring session to ensure that they were making progress, even if they could not meet as frequently. Krishna also encouraged Arjun to prioritize his leadership development.

KEY TAKEAWAYS

CHAPTER 6: THE PSYCHOLOGY OF MENTORSHIP: UNDERSTANDING THE DYNAMICS OF MENTOR-MENTEE RELATIONSHIPS

The Psychological Benefits of Mentorship for Mentees

- Empathy: Successful mentor-mentee relationships thrive on empathy, as it fosters trust and open communication.

- Motivation: Mentors should employ motivational tactics to inspire mentees to strive for personal growth and excellence.

- Cognitive Congruence: Aligning teaching methods with the mentee's learning style promotes effective learning experiences.

- Timely Feedback: Providing constructive and specific feedback enables mentees to reflect, improve, and take ownership of their growth.

- Two-way Learning: Mentors also benefit from the mentorship journey, gaining insights and refining their approaches.

- Psychological Benefits for Mentees: Mentorship offers psychological benefits such as increased self-confidence, better decision-making abilities, and overcoming self-doubt.

- Overcoming Professional Challenges: Mentors provide support and guidance to navigate difficulties in the mentee's career.

- Building Self-Confidence: Mentors help mentees recognize their strengths, address areas for improvement and develop self-confidence.

- Transformation through Mentorship: Real-life examples, like Stephanie's transformation, illustrate the positive impact of mentorship on personal and professional growth.

- Lasting Impact: Mentorship equips mentees with the tools and resources needed for future success, empowering them to achieve their goals with confidence.

The Psychological benefits of Mentorship for Mentors

- Overcoming Self-Doubt: Mentors may initially struggle with self-doubt and imposter syndrome, questioning their abilities to guide others. Seeking support from other mentors and recognizing their own strengths can help mentors overcome these doubts.

- Personal Growth and Development: Mentoring others can lead to personal growth and development for mentors. They can gain new perspectives, challenge themselves, and attend workshops and seminars to improve their coaching skills.

- Leadership and Coaching Skills: Mentors should continuously work on developing their leadership and coaching skills. This includes giving mentees autonomy, seeking feedback, and using effective coaching techniques to empower mentees.

- Building a Sense of Community: Mentors should foster a sense of community and belonging among their mentees. Organizing meetings, encouraging collaboration, and providing support can create a supportive environment for mentees.

- Finding Purpose and Fulfillment: Mentoring others can give mentors a sense of purpose and fulfilment in their work. It allows them to make a positive impact on the lives of others and find renewed passion for their own careers.

- Cultivating a Positive Impact and Legacy: Mentors can strive to leave a positive impact and legacy through their mentorship. By focusing on making a meaningful difference, inspiring others, and creating a culture of mentorship, mentors can leave a lasting legacy that continues to inspire future generations.

Common Mentor-Mentee Relationship Patterns and How to Navigate Them

- Importance of Boundaries: Maintaining professional boundaries while building a strong relationship with a mentee is crucial to avoid confusion and conflicts of interest.

- Power Dynamics: Mentors should create a safe space for open communications, allowing mentees to express their thoughts and opinions, despite the power dynamics that may exist.

- Conflict Resolutions: Mentors should be prepared to address conflicts with their mentees and work towards finding mutually beneficial resolutions.

- Diversity and Inclusivity: Mentors should understand and appreciate diversity, being mindful of differences that may impact their mentees' experiences to avoid breakdowns in communication.

- Accountability: Mentors should take ownership of their guidance, be accountable for their mistakes and actively work towards finding solutions to any issues that arise.

The Role of Trust in Mentorship Relationships

- Trust is essential: Trust forms the foundation of a successful mentorship relationship. It allows mentees to rely on their mentors for honest feedback, guidance, and support.

- Open communication: Trust enables open and transparent communication between mentors and mentees. Mentees should feel comfortable sharing their challenges, concerns, and aspirations with their mentors.

- Honest feedback: Trust allows mentors to provide honest and constructive feedback to mentees. Mentees should be receptive to feedback, even if it is difficult to hear, knowing that it comes from a place of genuine care and support.

- Reliance on mentor's expertise: Trusting the mentor's expertise and experience is important for mentees to fully engage in the learning and development process. It allows mentees to leverage their mentor's knowledge and insights to enhance their own skills and capabilities.

- Mutual support and best interests: Trust enables mentees to believe that their mentors have their best interests at heart. Mentors should demonstrate genuine care and provide support to help mentees grow and succeed.

- Valuable asset: A mentorship relationship built on trust becomes a valuable asset in a mentee's professional life. Mentees can turn to their mentors for guidance and support, knowing they have a trusted advisor by their side.

- Growth and development: Trust empower mentees to embrace growth and development opportunities. It allows them to take risks, explore new ideas and push beyond their comfort zones with the confidence that their mentor supports and believes in their potential.

- Long-term benefits: Building trust in a mentorship relationship yields long-term benefits. Mentees can continue to learn and benefit from their mentor's expertise even as they progress in their careers.

Common Barriers to Effective Mentorship Relationships

- Align expectations: Clear communication and alignment of expectations between mentors and mentees are crucial to avoid mismatched expectations and frustration. Discussing mentoring goals, approaches, and preferred styles can help establish a common understanding.

- Foster effective communication: Open and effective communication is vital for mentorship success. Mentees should feel comfortable reaching out to mentors for advice and support and mentors should actively encourage and provide guidance on communications skills.

- Overcome time constraints: Busy schedules can pose challenges to mentorship relationships. Both mentors and mentees should prioritize and commit to regular meetings, establish specific goals for each session and find suitable meeting times that accommodate their availability.

- Address power dynamics: Mentees may feel intimidated or hesitant to express themselves freely due to power dynamics or hierarchical structures.

Mentors should create a safe and open environment, actively encourage mentees to share their opinions and empower them to take ownership of their ideas.

- Embrace change and feedback: Mentees should be open to change and receptive to feedback to foster growth and development. Mentors can create a supportive environment where constructive criticism is encouraged and seen as an opportunity for improvement.

- Recognize and respect cultural differences: Mentors and mentees from diverse cultural backgrounds may face challenges in understanding each other's perspectives. Cultural sensitivity training and open discussions can help bridge the gap and foster mutual understanding and respect.

- Manage personality conflicts and communication styles: Differences in personality types and communication styles can lead to misunderstandings and conflicts. Mentors and mentees should invest time in understanding each other's preferences and adapt their communication styles accordingly.

- Build trust and respect: Trust and respect are essential for effective mentorship relationships. Mentors should provide guidance on earning trust, encourage mentees to be authentic and vulnerable and foster an environment where trust can flourish.

- Provide adequate training and resources: Mentees need access to proper training and resources to develop their skills effectively. Mentors can assist in identifying and providing relevant training programs and resources, and mentees should also proactively seek out opportunities for growth.

- Prioritize leadership development: Time constraints and competing priorities can hinder progress. Both mentors and mentees should prioritize leadership development and set aside dedicated time for mentoring sessions and personal growth.

To be continued in "The Power of Mentor" Volume-II

What is Covered in The Power of Mentor - Volume II

Chapter 7: Overcoming Mentorship Challenges: How to Navigate Conflicts and Overcome Obstacles in Mentorship

You'll discover how to navigate conflicts and overcome obstacles in your mentorship relationships. This chapter provides practical strategies for dealing with difficult mentors or mentees, the importance of clear communication, and addressing imbalances in the mentorship dynamic. Learn how to manage mentorship relationships when goals or priorities shift and understand the significance of setting boundaries to foster a healthy mentorship journey.

Chapter 8: The Power of Reverse Mentoring: How Learning from Younger Mentors Can Benefit You

This chapter highlights the immense value of reverse mentorship, where you learn from younger mentors. You'll explore the benefits of reverse mentorship for older professionals and discover how to find and connect with younger mentors. Learn effective strategies for leveraging the skills and knowledge of your younger mentors and address potential power imbalances to create a mutually beneficial and transformative relationship.

Chapter 9: Building a Stronger Network: How Mentorship Can Help You Expand Your Professional and Personal Connections

Uncover how mentorship can play a pivotal role in building a more diverse network. This chapter provides strategies for finding mentors outside of your immediate network and leveraging mentorship to build meaningful personal and professional connections. Embrace mentorship as a powerful tool to enhance your networking skills and etiquette, enabling you to create lasting relationships that foster personal and career growth.

Chapter 10: The Role of Mentors in Leadership Development: How Mentorship Can Help You Become a Better Leader

In this chapter, you'll learn how mentorship can significantly impact your leadership development journey. Discover how to find mentors who can help you develop specific leadership qualities and create a tailored mentorship plan for your leadership growth. Address common leadership development challenges and understand the crucial role of self-reflection in becoming a better leader.

Chapter 11: The Impact of Technology on Mentorship: How to Leverage Digital Tools and Platforms for Effective Mentorship

Explore the benefits of using technology to enhance your mentorship experience. This chapter outlines best practices for using digital communication tools and online platforms to find and connect with mentors. Overcome potential challenges in using technology for mentorship and get insights into the future of technology's influence on mentorship.

Chapter 12: Mentorship and Diversity: How to Find and Engage Mentors from Different Backgrounds and Perspectives

Discover the importance of diversity in mentorship relationships and how it can positively impact your personal and professional growth. This chapter provides strategies for finding mentors from different backgrounds and perspectives, building cross-cultural mentorship relationships, and overcoming potential barriers. Embrace the transformative power of diverse mentorship relationships.

Chapter 13: Mentorship Beyond Boundaries: How to Establish and Maintain Long-Distance Mentor-Mentee Relationships

This chapter emphasizes the benefits of long-distance mentorship relationships and provides practical strategies for finding and connecting with mentors outside your geographic area. You'll learn best practices for maintaining long-distance mentorship relationships using technology as a powerful tool. Address common challenges and discover how to foster a strong and supportive long-distance mentorship journey.

Through "The Power of Mentor" Volume II, you'll embark on a transformational journey that will equip you with the knowledge, strategies, and insights to maximize the benefits of mentorship in your personal and professional life. Embrace these powerful chapters, and I'm confident that you'll unlock your full potential as a transformational leader, fostering growth and success for yourself and those around you. Let's dive in and embrace the power of mentorship together!

May I ask you for a small favor?

I hope this note finds you well and filled with joy. As an author, I couldn't be more grateful for your decision to embark on this literary journey with me by reading my book, "**The Power of Mentor – Volume I.**" Your time and attention are incredibly valuable, and I am deeply touched that you chose to invest them in exploring the world I created within these pages.

Writing this book was a labor of love and knowing that it has touched your heart and mind means the world to me. Your support and encouragement have inspired me beyond measure, reminding me of the profound impact words can have on our lives and how they can forge genuine connections between strangers.

If it's not too much to ask, I would be profoundly grateful if you could take a few moments to share your thoughts on Amazon through a review. Your feedback will not only help potential readers discover the book but will also guide me in my future writing endeavours. Your honest opinion is priceless.

To put it straight – **Reviews are the lifeblood of any author.**

Additionally, I wholeheartedly recommend it. **"The Ultimate Leadership in You"** & **The Power of Mentor" Volume II** to anyone seeking to unlock their true leadership potential and make a profound impact on their life and the lives of those around them. Your personal recommendation could be the catalyst for transformation in the lives of those who need it most.

Remember, a heartfelt review has the power to touch the lives of countless readers, guiding them to a story that could make a difference in their lives too.

Once again, thank you for joining me on this literary adventure. Your presence in my journey is a cherished gift, and I am forever grateful for the opportunity to connect with you through my words.

Cheers,

Sreekanth Ganeshi

Books By This Author

"The Ultimate Leadership in You"

Are you ready to unleash the leader within and transform your life? "The Ultimate Leadership In You" is a compelling guide tailored for aspiring leaders like you, empowering you to cultivate powerful leadership skills, lead with integrity, and create a lasting impact on the world around you.

Discover the transformational power of your leadership potential as you embark on a journey of self-discovery and growth with **"The Ultimate Leadership In You."**

Here is what you can get from this book:

Personalized Leadership Development: This book is tailored for you, providing step-by-step guidance and practical insights to cultivate your unique leadership abilities.

Understanding Your Why: Unravel the core motivations behind your desire to lead and take the first step towards becoming a strong and confident leader.

Lead Yourself First: Master the art of leading yourself first, cultivating self-awareness, resilience, and a growth mindset, laying the foundation for exceptional leadership.

Five Proven Strategies for Leadership Mastery: Discover five powerful ways to develop and enhance your leadership skills, empowering you to inspire and influence others with purpose and conviction.

Embrace the Servant Leader: Embrace the transformative concept of servant leadership, where you prioritize the needs of your team and foster a culture of trust and collaboration.

Integrity: The Keystone of Leadership: Unveil the most important quality of a leader—integrity, and witness how it fosters unwavering trust and credibility among your peers.

Vision: The Driving Force of Achievement: Harness the incredible quality of a leader—vision, inspiring your team to reach new heights and achieve shared goals.

Build Meaningful Connections: Become a good connector, forging strong relationships that elevate team performance and foster a positive work environment.

The Power of Your People: Recognize that your team is your most valuable asset, and learn how to unlock their potential, driving collective success.

Tap into your true leadership potential, gain indispensable insights, and accelerate your success with "The Ultimate Leadership In You." Embrace the life-changing journey of personal growth, empowerment, and achievement. Unleash the leader within and become a beacon of inspiration for others. Empower yourself with the wisdom to transform lives and leave an enduring legacy.

The time for transformation is now! Seize this life-changing opportunity and unlock your leadership potential with "The Ultimate Leadership In You**." Click "Buy Now"** to embark on this empowering journey and embrace the profound impact of your transformed leadership on yourself and those you lead. Don't hesitate; take action and shape your future as a remarkable leader!

Books by this Author

Direct Link to Buy Now

Follow us on Social Media

https://www.facebook.com/SreekanthGaneshi

https://www.facebook.com/groups/sreekanthganeshi

https://twitter.com/1Sreekanth_G

https://www.linkedin.com/in/sreekanthganeshi/

https://www.instagram.com/sreekanthganeshi/

Notes

Printed in Poland
by Amazon Fulfillment
Poland Sp. z o.o., Wrocław